BEYOND COMPLIANCE

Proven Methods for Implementing ***Effective*** **S**tatistical **P**rocess **C**ontrol Charts

A Practical Guide for Transforming "SPC for Show" into "SPC for Improvement" to Achieve Operational Excellence

By

Steven F. Bergeron

Quality • Lean • Reliability • Focused

First Edition

ISBN: 979-8-99475599-0-5 (Paperback)

Library of Congress Control Number: 2026906152

Published by QLRFocus Publishing

Holden, MA

Printed in the United States of America

Charts, figures, and examples in this book are reconstructed or anonymized for educational purposes. No proprietary or confidential company data is disclosed.

TABLE OF CONTENTS

PREFACE

Welcome to a fresh, practical approach to Statistical Process Control (SPC). This book is designed for practitioners and leaders who are ready to move beyond superficial, compliance-driven SPC practices and toward systems that genuinely improve process performance. Drawing on more than four decades of quality and reliability assurance leadership, I present not only the underlying theory, but proven, actionable methods for implementing, evaluating, and sustaining SPC charts and programs in real-world manufacturing and supplier-management environments.

At the heart of this book is a simple but powerful theme: transforming "SPC for Show" into "SPC for Improvement." Organizations striving for operational excellence – not merely passing audits – will find clear guidance on how to diagnose the current state of their SPC systems and how to proceed with a meaningful transformation. For new implementations, the book provides a structured roadmap that helps teams avoid the common pitfalls that undermine SPC Programs before they ever become useful.

Real-world case studies and examples ground the concepts in practical reality, making the material accessible and relevant for engineers, auditors, managers, and executives alike. The book addresses common pitfalls, such as incorrect control chart setup, competing priorities, and the subtle *"domino effect"* that can quietly erode SPC programs over time. You will find detailed instructions on applying the innovative Effective Sample Size technique, best practices for control chart qualification, combining equivalent equipment, and practical advice on interpreting and using capability indices. Additional chapters provide guidance on when and how to calculate realistic product specifications, methods for auditing SPC charts and programs, and a highly effective Maverick Lot Program for managing Ship-To-Control requirements as well as protecting your organization during your SPC journey.

Throughout the book, I emphasize how successful SPC Programs are directly connected to business performance. Unstable processes drive scrap, rework, overtime, warranty exposure, and unpredictable throughput. Misinterpreted data leads to over-adjustment, excess inventory, and inefficient capital allocation. Conversely stable, and capable processes deliver measurable economic benefits, reduces costs of poor quality, improves yield, enables predictable cycle times, and delivers more reliable products.

This book bridges the gap between compliance and true process improvement. It provides the tools, mindset, and roadmap needed to build SPC systems that deliver lasting quality, reliability, and financial performance.

From experience I can confidently say that *Effective* SPC Programs are a direct contributor to profitability and competitive advantage.

By the end of this book, you will understand not only how to implement SPC correctly, but how to establish a culture and system where SPC continuously drives business results.

– Steven F. Bergeron

1. INTRODUCTION

There are many books that describe how to set up and use Statistical Process Control (SPC) charts. I have used many of them, learned from them; and certainly, recommend them. What is different about this book? This book describes *proven methods* on how to *effectively* implement Statistical Process Control (SPC) Charts for *variables* in your manufacturing facilities as well as in your supplier's facilities. This book also describes how to assess already implemented *variable* control charts to determine their effectiveness, and if judged *ineffective*, how to fix them.

This book is based on my firsthand experience of how I learned to *effectively* implement Statistical Process Control (SPC) Charts in production beginning when I was a Process Engineer through my experience as a Chief Quality Executive and finally as a Consultant. I address all the practices that I experienced during my career that interfered with effective SPC Charts. Indeed, I provide methods and approaches to address and prevent these bad practices. The good news is that everything I discuss and methods that I propose are consistent with accepted SPC practices. I am also introducing new methods and techniques that will change "SPC for Show" programs to SPC programs that really work.

In addition, I discuss key issues routinely faced by engineers and implementors, such as: Can your process meet specifications now? Is it likely your process will meet specifications in the future? Given the current process, what should your specification be? How can you make sure your suppliers do not ship you bad products even as you are trying to get them to *effectively* implement SPC?

The target audience for this book are SPC owners and implementors, SPC auditors and Management. Quality, Product and Process engineers are typically responsible for the setup and implementation of the initial SPC Charts as well as ensuring continued SPC Chart effectiveness. Auditors (e.g., Quality Auditors or Supplier Engineers) are typically responsible for ensuring that SPC Charts within the company or supplier's SPC Charts are effective and that they follow adopted standards and/or comply with customer requirements. Of course, I cannot ignore management. Once Management understands how to judge the effectiveness of the company's SPC program, they are more likely to support the described practices. Even more important, they are less likely to take actions or create policies that interfere with SPC effectiveness.

I began my career in 1978 as a Process Development Engineer. As a Process Engineer, I learned to use statistical techniques, such as Design of Experiments, to help me develop processes. Once processes were developed, I had to transfer and qualify these processes in manufacturing. This is when I first took training and learned to use SPC Charts to qualify, control, stabilize and improve my processes. My company took notice of my consistent success and asked me to join the Quality Team to assist with the deployment of these techniques throughout the company.

My quality skills and certifications within the Quality profession grew throughout my career. I am proud to say that I became a Certified Quality Manager (CQM), Certified Six-Sigma Black Belt, Certified Auditor for ISO 9001, ISO 17025, and TS 16949 Quality Standards, and a Certified Lean Manufacturing Facilitator. I trained and implemented Reliability and Advanced Product Quality Planning (APQP) techniques such as PPAP, FMEA, MSA and of course SPC. I was twice appointed by the US Dept of Commerce to be on the Malcolm Baldrige National Quality Award Board of Examiners. I guided multiple companies, as their chief Quality Executive, to achieve and maintain certifications to ISO 9001, ISO 13485, IATF 16949, AS 9100, and ISO 14001 and led their quality improvement journeys.

As my career grew, I was recruited to other companies to assess, build, re-build and/or modernize company-wide Quality Management Systems to meet business goals, adopted Quality Standards such as those mentioned above, or other customer requirements. I take considerable pride that I was promoted to my first Quality Executive position as Vice President of Quality and Reliability Assurance by the time I was 40.

As the company's Chief Quality Executive, I was responsible for all elements of the company's Quality Management System. However, this book focusses only on the quality sciences associated with Statistical Process Controls. In **all** the companies that I worked for, I initially found ineffective or non-existent SPC programs. Even though I was the company's Chief Quality Executive responsible for the total QMS, I found myself personally teaching and coaching company personnel, including my own Quality organization, and our suppliers, on how to implement *effective* SPC programs using the methods described in this book. I founded, as President, QLRFocus, LLC

(Quality Lean Reliability Focus) where I supported startups, small enterprises, and global corporations in advancing operational excellence through specialized Quality, Lean, and Reliability strategies, and of course, Statistical Process Control.

All the companies I worked for offered products that were well received in the marketplace. Yet the success of these companies, and their products, depended on their ability to implement and maintain high volume manufacturing facilities that delivered products of exceptional quality along with high reliability at low cost as demanded by their customers and the marketplace. Leading companies adopted proven quality methods, without prompting, to help them achieve their business goals by *attempting* to implement these methods into their production facilities. Others adopted advanced quality techniques only when demanded by their customers or when required by standards adopted by the targeted industries.

Examples of popular and proven quality practices include, but are not limited to such techniques as Design of Experiments, Reliability Prediction and Improvement, Six Sigma, APQP, Lean Manufacturing and of course Statistical Process Controls (SPC) ← the focus of this book. *Ineffective* SPC Chart implementation kept production costs high, leading to high scrap rates which led to poor margins, poor product quality, and high customer complaints which led to expensive warranty claims. Because of this, some of these companies had to exit the Automotive and Aerospace markets. The opposite is also true, *effective* SPC Chart implementation led to market dominance as scrap rates achieved record lows resulting in high margins, record low customer complaints which led to minimal warranty claims. Business expanded!

As previously mentioned, throughout my career I found that upon joining a company and after my initial assessment, their SPC programs did not deliver the intended results. This was true for both the companies I worked for and for the suppliers of companies I worked for. I label these programs "SPC for Show" meaning that the organizations were able to show customers, auditors, and/or internal reviewers that a SPC program had been implemented. Their implementation satisfied requirements and Quality Standards such as ISO 9001, the automotive standard IATF 16949 (QS 9000) and AIAG Core Tool Standards, Aerospace standard AS 9100, ISO 13485 and other customer requirements. Indeed, you can audit and observe implemented SPC programs as they are often using production software –

which makes it easy to see SPC charts. But what I found is that most assessments and audits could only verify the existence of SPC programs. They could not determine whether the SPC programs delivered the promised results. This issue is particularly problematic for those companies that rely on outsourced manufacturing.

The primary problem is that most implementors of SPC programs do not recognize that their implementation is not really working even though most of them took SPC training, and they were just trying to implement the training. My experience indicates that standalone SPC training seemed to follow sound statistical theory yet were incomplete as the training did not address typical manufacturing practices. These practices, which I discuss in detail in this book, interfere with *effective* SPC Charts. I also found that most of the training was linked to teaching engineers how to use the SPC production software program, rather than how to effectively setup a control chart. This type of training provides *some* statistically based SPC theory yet was primarily focused on how to use the SPC production software program. I also consistently found that some software program features violated statistical principles or created bad practices yet were included in the production software because users asked for them. In both cases, manufacturing practices and/or some software features drove improper actions or behaviors of manufacturing personnel. Since they were not addressed in training or were allowed by the software program, the implementors believed they were acceptable. However, they caused the SPC Chart to become ineffective. In most cases, these behaviors reduced process capabilities, increased out of control events, and caused higher scrap rates.

So, my challenge was to learn how to lead the organization to fix their SPC programs and reap the benefits of effective SPC Charts. Adoption rates varied, but I can confidently say from experience: those individual implementors (e.g., quality, product, product, and supplier engineers) who implemented these approaches and methods were rewarded with very stable processes, improved yields, less scrap - all the benefits of effective SPC Charts! For those organizations that deployed these methods and approaches across their factory achieved market dominance with record low scrap rates, high margins, and minimal warranty claims.

This book discusses real case studies and examples that I addressed during my career. However, I take great care not to mention the company I was working for to protect the privacy of each company. In addition, the charts, graphs, and data used are re-constructed to protect the privacy of company data.

The methods and approaches described in this book are consistent with accepted statistical methods and Quality Standard requirements such as ISO 9001, IATF 16949, ISO 13485, AIAG Core Tool Standards and AS9100 (i.e., if implemented as described, they will pass any Quality Management System (QMS) Audits associated with these standards).

After retiring, I reflected on my work as Chief Quality Officer and consultant, where I found the companies' SPC Programs were ineffective, serving mainly as "SPC for Show." I also experienced their remarkable business performance improvements as they transitioned their SPC Programs from "SPC for Show" to "SPC for Improvement" as they implemented the methods described in this book. I authored this book to share the techniques that helped my teams succeed, so other professionals and companies can achieve similar results.

Ok – Let's get started!

2. CHAPTER SUMMARIES

Here I provide a quick overview for each Chapter in this Book. The goal is to provide you, the reader, with a summary of each chapter, so you may quickly find a chapter of interest.

Chapter 3: SPC Quick Overview

This chapter introduces Statistical Process Control (SPC) as a set of methods for monitoring, controlling, and improving processes through statistical analysis. SPC is highlighted as essential for ensuring quality, reducing defects, saving costs, enabling early problem detection, supporting data-driven decisions, increasing customer satisfaction, ensuring regulatory compliance, and fostering continuous improvement.

The chapter credits Walter A. Shewhart for developing control charts in the 1920s, which distinguish between "common causes" (natural) and "special causes" (assignable) variations in processes. Control charts visually identify whether a process remains stable and in control or if it has shifted due to special causes, prompting investigation and corrective action.

Key SPC terminology and definitions are provided, including types of data, critical-to-quality characteristics (CTQs), population and sample concepts, control chart out-of-control rules and action plans for out-of-control (OOC) and out-of-specification (OOS) events. The chapter explains the structure of a typical control chart, including centerlines, upper/lower control limits (usually set at ±3 standard deviations), and the use of zones and special cause rules to detect unusual process behavior.

Several common types of control charts for variable data are described, with recommendations to primarily use three types: Individuals & Moving Range (X_i & MR), X-bar & Range ($\bar{X}$ & R), and X-bar & Standard Deviation ($\bar{X}$ & s) charts. The chapter outlines the formulas and constants used for these charts, noting that most software automates these calculations.

The chapter concludes by addressing common pitfalls that render SPC charts ineffective, such as improper setup and competing priorities. These issues can lead to frequent false alarms, ineffective corrective actions, and a shift from genuine process improvement to mere compliance ("SPC for Show"). Subsequent chapters address, in detail, these pitfalls and provide solutions to ensure SPC drives true process stability and improvement.

Chapter 4: Competing Priorities: Product or Process Control

This chapter addresses the most important factors contributing to ineffective SPC programs: the ongoing challenge between Product Control and Process Control. The discussion begins with an overview of the distinct objectives, relevant statistical tools, principal strategies, and desired outcomes associated with Product and Process Control.

It examines Product Control's use of AQL statistical methods for assessing whether randomly sampled production lots meet Acceptable Quality Levels required for shipment, contrasting these with Process Control's SPC techniques aimed at continuously reducing variation for critical-to-quality (CTQ) characteristics.

Further, the chapter explores how these competing priorities are supposed to work together by comparing how limits are established, and what actions should be taken when a sample is found to be Out-of-Control (OOC), or if the sample contains Out-of-Specification (OOS) units.

The chapter introduces the concept that conflicts may arise when organizations seek to take additional measures beyond product containment upon detection of OOS samples, or when practices are influenced by customer "Ship-to-Control" requirements.

The chapter also explores the potential detriment of introducing specifications for Process CTQs when they are not needed as another example of well-intentioned actions that lead to ineffective Control Charts.

The Chapter ends with 3 key recommendations: (1) whenever possible Separate Product CTQs from Process CTQs; (2) don't invent specifications that are not needed; and, (3) if you need to monitor a Product CTQ be sure OCAPs address process control and the OSAPs are only used to contain and disposition OOS material.

Chapter 5: Top Reasons Why Control Charts Are Setup Wrong

This chapter discusses the Top Five Reasons Why Control Charts are set up wrong and their impact on Control Chart Effectiveness. They are:

1. Control Charts Monitor Production Lots Not Process Runs.

   The chapter discusses the high risk of OOC event errors if care is not taken to ensure the data plotted on a control chart represents process runs and not production lots.

2. Specification Limits for Individuals are Placed on Control Charts for Averages.

   When spec limits were placed on control charts, the nature of OCAPs usually changes. Often OOC events are ignored if the average was found to be in spec. OCAP actions often evolved to only act when OOS events occurred. This behavior was often caused by customer requirements as they demanded spec limits be placed on Product CTQ control charts. The reasons why you should not do this are discussed. Simply stated: do not place specification limits for individuals on Control Charts are for averages. This approach frequently leads to degraded process performance as the Control Chart becomes ineffective.

3. The Control Limits are Setup Using the Wrong Sample Size.

   The chapter explains how control limits are calculated wrong even though the calculations use the actual sample size taken. It explains that control limit calculations assume samples drawn from each process run are random and independent from each other. However, often selected samples may be correlated with each other, which is not an obvious situation. When this occurs, it causes the control limits to be tighter than they should be, which causes false OOC events, then Process Owner frustration which ultimately triggers the *"domino effect"* leading to an ineffective Control Chart. The chapter introduces the *Effective Sample Size* technique to address this problem which is described in Chapter 6 in detail.

4. Evidence of a Lurking Variable is not Addressed During Setup

   Often Control Chart setup data indicates the presence of a *Lurking Variable.* The chapter discusses techniques on how to detect *Lurking Variables* during control chart setup. It also discusses the most common *lurking* variables: measurement error, control chart plots production lots and not process runs, inconsistent process execution, inconsistent process adjustments, and that equivalent process equipment are often plotted on the same chart too early. The chapter discusses how to detect or prevent these during control chart setup. It introduces the concept of "Copy Exact" and "Copy Similar" equipment, i.e., equivalent, or similar equipment that are supposed to deliver the same process. Often, they are too quickly combined on the same control chart. It points to Chapter 8 on techniques for effectively combining equivalent equipment. If *Lurking Variables* are ignored, the control chart can be ineffective and trigger the *"domino effect."*

5. The Good, The Bad and The Ugly OCAPs

   This chapter begins by describing in detail real examples of OCAPs triggered by OOC events which demonstrate the effect of different OCAP strategies. The chapter then describes various OCAP strategies and from experience, and the long-term effects for each OCAP strategy. The Best OCAP strategy is to stop the process, investigate the reason for the process shift and then remove the reason for the OOC event. This type of OCAP consistently reduced OOC events, improved process capability, and stability. Good OCAPs usually maintain process capability and stability include consistent and/or known and proven process adjustments. Bad OCAPs include actions only taken when an OOS event occurred while ignoring OOC events. Ugly OCAPs included allowing on-demand (such as Call Engineering) unproven adjustments. These OCAPs most often degraded process capability and stability and often were the root cause for yield busts. The chapter describes how Ugly OCAPs often invite specification control versus process control.

The chapter makes the case that even if you setup a control chart correctly, the type of OCAPs defined for a OOC event also make a huge difference with respect to control chart effectiveness.

Chapter 6: Discovering the Effective Sample Size (n_{eff})

This chapter describes the exact case where I discovered that the sample size that was taken during a control chart setup led to incorrect control limits when applying the control limit calculation rules, i.e. the control limits were just too tight to make the control chart effective.

It describes a semiconductor wet etch process where 18 samples consisting of 6 wafers per process run, with 3 sites sampled per wafer for a total of n =18 for each process run. The chapter describes how traditional control limit calculations led to 60% of the setup runs violating Rule #1 (1 point out-of-control) when using n =18 (6 wafers x 3 sites per wafer) in the control limit calculations. The Process Engineer (Control Chart Owner) was concerned that she would not have enough time to stop the process 60% of the time to launch an investigation to determine the special cause. In addition, she and other engineers agreed, including me, that the process was not that bad.

The chapter describes how we reviewed her calculations and verified that she applied the formulas correctly. This caused additional review of core principles of SPC and the *Central Limit Theorem*. One core principle emerged: the Central Limit Theorem assumes that samples taken (and used to calculate control limits) are random and independent. Our investigation explored whether this assumption was true. The chapter describes the breakthrough moment when key equations were used to solve for n rather than assume n =18. The chapter further describes how we used correlation analysis to determine the most uncorrelated (most random) samples from the original sampling plan. After recreating the control chart using the calculated n = 2, the control chart appeared *well-behaved* and was implemented in the factory. The case study then describes how we were able to link the calculated n to physical characteristics of the process. I ultimately labelled this approach the *Effective Sample Size* (n_{eff}) method and used it throughout the rest of my career.

The chapter further explains how I consistently linked the calculated *Effective Sample Size* (n_{eff}) with a process's physical characteristics across various processes, equipment types, companies and in different industries.

The chapter ends with key conclusions: Setting up a control chart based on CTQ samples that are correlated with each other (i.e., not random, or independent) will lead to control limits that are too tight. The proposed solution, use the *Effective Sample Size* (n_{eff}) method to determine the most uncorrelated sample size to calculate control limits.

Chapter 7: How to Setup and Qualify a Control Chart

The previous chapters covered key issues that I discovered led to ineffective SPC Control Charts and unsuccessful SPC Programs. This chapter shifts the focus to addressing these findings and outlines a clear 16-step process for setting up and *qualifying* an effective control chart for variables. These steps include all traditional practices and include my methods for addressing identified key issues.

The 16 steps discussed in this chapter on how to *setup* and *qualify* a control chart *before* it is released to production are:

1. Identify the Process, Process Equipment, and the CTQ.
2. Verify the Readiness of Measurement Equipment.
3. Verify Work Instructions and Production Operators are Certified.
4. Define the Initial Sampling Plan (n_{init}).
5. Setup Trial Control Limits for Pilot Runs.
6. Conduct 30 Process Runs and Measure the CTQ.
7. Analyze the Dataset for the Effective Sample Size (n_{eff}).
8. Determine the Most Random and Independent Samples.
9. Decide if the Ongoing Sampling plan Should be Reduced to (n_{eff}).
10. Construct a Control Chart using (n_{eff}) to set Control Limits.
11. Analyze and Remove Obvious Lurking Variables.
12. Calculate the *Stability Index* (S_i) and Apply the 85% Rule.
13. Define Quick Response OCAPs.
14. Define OSAPs for Product CTQs (when necessary).
15. Enable Out-of-Control (OOC) Rules with Quick Notifications.
16. Release the Qualified Control Chart to Production.

Chapter 8: Combining Equivalent Processing Equipment Control Charts

This chapter addresses a very common situation in most factories. Most factories have multiple pieces of equipment designed to deliver equivalent processes. Some of the equipment are "Copy Exact," meaning the alternate equipment is just another identical piece of equipment (for example the exact same model) configured to deliver the same process. Others are "Copy Similar" meaning the alternate equipment is a different piece of equipment configured to deliver the same process.

The method describes how to determine if "Copy Exact" or "Copy Similar" process equipment could and/or should be added to a previously *Qualified Control Chart.*

The chapter describes situations when "Copy Exact" equipment can use the same n_{eff} as the original equipment used for the *Qualified Control Chart*. However, it emphasizes the need to re-calculate n_{eff} for "Copy Similar" equipment.

The method indicates that you should conduct 8 consecutive process runs using the candidate equipment ("Copy Exact" or "Copy Similar"). Plot the 8 process runs on the *Qualified Control Chart* and then apply Special Cause Rules 1 and 4.

If the candidate equipment passes Special Cause Rules 1 and 4, the candidate equipment is judged to be delivering statistically equivalent process variation. In this case, the candidate equipment can be combined on the *Qualified Control Chart.*

However, if a violation is detected, it means that the candidate equipment is not delivering an equivalent process and should not be plotted on the *Qualified Control Chart.* The chapter suggests 2 options: (1) investigate and remove the special cause, then conduct 8 more runs and repeat the analysis; or (2) plot the process runs on a separate control chart until statistically equivalent conditions emerge.

Combining equipment before proving their equivalence leads to numerous issues, such as competing priorities and ineffective OCAPs (adjustments, engineering calls, etc.). This situation significantly increases workload as the organization deals with false OOC events, ultimately resulting in ineffective control charts.

Chapter 9: When to Recalculate Control Limits

This chapter outlines situations when control limits should be recalculated, including: *(1) after identifying* and removing special causes; (2) following a deliberate process adjustment; and (3) after specified review periods for a well-behaved process. It also provides methods for recalculating control limits.

For example, in Case 1, after investigating and applying corrective measures to address a special cause, the recommendation is to conduct 10 additional process runs on the affected equipment and apply Rule 4 without recalculating control limits. If no further violations occur, an additional 20 runs are conducted for a total of 30, at which point control limits are recalculated.

Similar procedures are recommended for Case 2 following a deliberate process adjustment intended to shift the process average.

The chapter further describes how and when to recalculate control limits after defined time intervals, suggesting the use of the 85% Rule to help determine if recalculation at set intervals is warranted. If the 85% Rule is not met, the focus should remain on investigating special causes rather than recalculating control limits.

Recalculating control limits for *well-behaved processes* using these criteria enhances the accuracy of control limits, aligns the control chart with current process performance, helps identify changes in process behavior, and supports compliance with quality standards or customer requirements.

The chapter concludes that recalculating control limits at appropriate intervals supports ongoing reliability of control charts for monitoring and maintaining process stability and quality. Adjusting control limits as the process evolves contributes to improved process stability and capability.

Chapter 10: Capability and Performance indices

The chapter introduces and explores the process capability indices (C_p, C_{pk}) and the process performance indices (P_p, P_{pk}).

First, the Chapter examines key reasons why these indices are important: including the ability to consistently evaluate and compare capabilities across different processes within an organization, across redundant equipment, suppliers and across industries; assisting management in making informed decisions about process control; and helping organizations demonstrate compliance to internal or customer requirements.

The discussion then provides key formulas, definitions, and assumptions for both process capability indices (C_p, C_{pk}) and the process performance indices (P_p, P_{pk}). It also suggests the use of the 85% Rule to determine if the process is sufficiently stable to calculate (C_p, C_{pk}), or, if not sufficiently stable suggests using (P_p, P_{pk}) performance indices instead.

The chapter also explores how customer C_{pk} requirements can initiate the "*domino effect*" leading to "SPC for Show" programs. The chapter then shifts to other common C_{pk} use errors including reporting C_{pk} for individual process runs; how to correctly take samples to estimate C_{pk}; and advises when you should be reporting P_{pk} instead of C_{pk}.

The chapter ends with a recap of recommendations including: using the *Stability Index* ≥ *.85* to determine if you should report Capability Indices (C_p, C_{pk}) or Performance Indices (P_p,P_{pk}); don't invent product specifications for Process CTQs; don't report C_{pk} for individual process runs unless you measure all units within the process run; and, report P_{pk} when combining datasets from various equipment unless the *Stability Index* ≥.85.

Chapter 11: Calculating Product Specifications

This chapter explores techniques to calculate product specifications using capability or performance indices. The chapter explains how to use these methods for both double-sided and single-sided specifications.

Specifically, the Chapter explains how to use C_{pk} formulas to calculate potential specifications from *qualified* control charts for Product CTQs. The methods essentially rearrange the C_{pk} formulas to solve for specifications (USL and LSL). Since the process is reasonably stable (*Stability Index* ≥.85), this provides you with BEST-YOU-CAN-DO specification estimate.

In situations when your process is not sufficiently stable (*Stability Index* < .85); or you wish to combine all equivalent equipment before they can be combined on a single control chart (see Chapter 8) it recommends using re-arranged P_{pk} formulas. When you use this method, the candidate specifications BEST-MATCH-YOUR-SHIPMENTS with your current equipment set without assuming further improvements.

These are useful techniques to adjust product specifications to actual process performance. My experience suggests that Design Engineering welcomes this analysis and adjusts product specifications about 80% of the time. These methods are particularly useful to both improve current factory performance, improve customer acceptance of products as well as influence specification setting for the next generation products. Experience also suggests successfully using these methods often changed production's focus by eliminating factors causing the *"domino effect."*

Chapter 12: Auditing SPC Programs for Effectiveness

Assessing the effectiveness of SPC programs is essential to ensure they achieve their intended objectives and add value to the company. This chapter pulls all the learning's from previous chapters and provides a method for auditing the effectiveness of individual control charts. When multiple control charts are audited using these methods, the results can determine the overall effectiveness of the organization's SPC Program. These methods can be used to audit your organization or audit the SPC Program of your suppliers.

The chapter begins by providing an individual control chart scorecard consisting of 15 criteria. Use each criterion during the audit on each control chart to determine whether the control chart has either implemented best practices described in this book; or is adversely affected by known practices that will interfere with the control chart effectiveness. An adjustment + or - points is added or subtracted from the starting score to achieve a final control chart score. The chapter explains each criterion in detail and provides examples of criteria as judged necessary.

The next section in the Chapter provides interpretation of individual control chart scores ranging from 200, the highest and most effective control chart to 0 the lowest score and most ineffective. It also provides guidance on what to expect from a control chart for a given score. Examples with appropriate interpretations are also provided.

The Chapter discussion then shifts and provides guidance on how to combine results from various control charts for an organization. The discussion suggests auditing at least 20 different control charts. It recommends selecting both Product and Process CTQs as well as sample different process owners and department owners. Once you have audited 20 or more charts, you have enough data to pass judgement on whether SPC Program is very effective, moderately effective or failed too many criteria and be judged ineffective. Guidance is provided on what scores constitute passing the audit as well as what scores indicate a failed audit.

While these methods help audit an SPC Program and judge its outcome, their main value lies in using the audit data effectively. Reviewing findings with management and control chart owners prompts them to address issues, leading to better process stability, higher yields, consistent product quality, and lower costs–the core of the *SPC Promise*.

Chapter 13: The Most Effective Maverick Lot Program

This chapter highlights the practical applications and effectiveness of the Maverick Lot Programs in manufacturing environments, particularly in situations requiring "Ship-to-Control" compliance, during the rollout of Statistical Process Control (SPC) programs, and when addressing findings from a failed SPC audit. I share personal experiences demonstrating how this program protects organizations and improves product reliability.

The chapter begins by defining a "Maverick Lot" as a production batch with statistically significant deviations in critical-to-quality (CTQ) parameters, even if all units meet specified limits. It distinguishes between Maverick Lots (lot-level outliers) and Maverick Units (individual product outliers), stressing that these can signal latent process issues not caught by standard tests. Two industry standards, JEDEC JESD50C and AEC-Q004, are introduced as frameworks for managing outlier lots and units, emphasizing proactive defect prevention and continuous improvement.

Implementation challenges are discussed, warning that applying Maverick Lot Programs to multiple CTQs throughout production can undermine SPC effectiveness by creating competing priorities. Instead, the chapter recommends focusing on Final Test Yield monitoring using Individuals/Moving Range (I/MR) control charts, as allowed by the JESD50C standard. This approach enables detection of abnormal lots (Maverick Lots) without disrupting SPC processes.

A comparison table illustrates the difference between Final Test Yield (lot-level focus) and Part Average Testing (PAT, unit-level focus), noting that PAT is often impractical in high-volume manufacturing and can lead to operational delays.

The chapter further describes how monitoring Final Test Yield with I/MR control charts allows for effective containment and investigation of abnormal lots, resulting in yield improvements and reduced customer complaints. Automation is highlighted as key to enforcing Maverick Lot Program rules and ensuring prompt responses to out-of-control events.

Case studies of two suppliers demonstrate the program's impact. Supplier A, which adopted the Maverick Lot Program as a shipping criterion, saw significant yield and reliability improvements after implementing the program alongside SPC enhancements. Supplier B, initially resistant, eventually accepted the program after seeing comparative performance and financial benefits, leading to similar improvements.

The chapter concludes that a well-implemented Maverick Lot Program, especially when coordinated with effective SPC practices, fulfills stringent customer requirements, protects organizations during SPC Program improvement efforts, and drives higher reliability and lower defect rates across industries such as automotive, aerospace, and semiconductors.

3. SPC QUICK OVERVIEW

Statistical Process Control (SPC) methods are used to monitor, control, and ideally improve processes using statistical analysis. Implementing a Statistical Process Control (SPC) program can bring significant benefits, particularly in manufacturing, service industries, and other quality-driven environments. Here are some primary reasons for adopting such a program:

1. **Improved Quality Control**: SPC helps monitor and control processes to ensure consistent product quality and reduce product variability.
2. **Defect Reduction**: By identifying and addressing sources of process variation, SPC reduces defects (including Out-Of-Specification (OOS)) Material) and waste, leading to more efficient operations.
3. **Cost Savings**: Fewer defects mean reduced costs associated with rework, scrap, and customer complaints, ultimately improving profitability.
4. **Early Problem Detection**: SPC tools allow for real-time monitoring, enabling the identification of issues before they escalate.
5. **Data-Driven Decisions**: SPC relies on statistical data, providing a factual basis for decision-making and process improvements.
6. **Increased Customer Satisfaction**: Producing high-quality, reliable products or services builds trust and loyalty among customers.
7. **Regulatory Compliance**: SPC can aid in meeting industry standards and regulatory requirements by maintaining consistent and traceable quality levels.
8. **Continuous Improvement**: SPC fosters a culture of ongoing process improvement, supporting Lean and Six Sigma initiatives.

Walter A. Shewhart is credited with developing the first control charts in the 1920s while he was working at Bell Laboratories. Shewhart studied measurable characteristics of processes and concluded that stable manufacturing processes display natural variation, which he labelled "common causes" of variation. Occasionally processes display unnatural variation indicating that the process shifted away "common causes" of variation and its historical performance. He labelled the source of this unnatural variation as "special or assignable" causes of variation.

He developed statistical methods and graphical representations of these methods and called them Control Charts. These Control Charts can be used in/by manufacturing to monitor the stability of a process and quickly detect the presence of these special causes. The detection of a special cause indicates that the process has shifted and is no longer delivering its *historical performance*. By investigating and removing the reason(s) for the special cause, the process will return the process to its historical performance. If the special cause was hidden within the previous distribution by virtue of chance, locating and removing the special cause will remove that component of variation, which reduces the overall distribution - thereby improving the variability of the process!

Before we get to studying the characteristics of a typical control chart, we need some terms and definitions. Table 1 provides some key definitions pertinent to SPC Control Charts. These definitions are used throughout the book.

Table 1 Key Definitions

Term	**Definition**
Variable Data	• A quantifiable and measurable characteristic that can be measured on a continuous scale and represented with a numerical number.
Critical-To-Quality Characteristic (CTQ)	• A Key measurable characteristic of a product or process whose performance is monitored and controlled to meet company and/or customer product/process quality requirements.
Process CTQ	• Process CTQs are the critical aspects of a process that must be controlled to ensure the process consistently produces the desired output. • They should not have specification limits. • Examples: Etch rate, Deposition rate. • Control charts for Process CTQs monitor the stability and performance of the process over time.
Product CTQ	• Product CTQs are the critical characteristics of a product that must be met to satisfy product requirements. • They should have specification limits. • Examples: Thickness, Voltage. • Control charts for Product CTQs monitor the stability, quality, and consistency of delivering product characteristics. • (Note: Some customers allow categorizing Product CTQs into three classifications: Critical, Key, and Significant. Refer to Chapter 10 for more information.)

Term	Definition
Population	• A sufficiently large collection of CTQ data that represents the complete set of possible CTQ values over time.
Statistical Frequency Distribution	• The distribution of CTQ values indicates the relative frequency of each possible CTQ value that occurs over time. Most often plotted as a histogram.
Population Parameter	• A summary statistic depicted using Greek symbols, which describes the distribution of a mathematically modeled population of CTQ values for variable data. • The population parameter for average is $\boldsymbol{\mu}$. • The population parameter for standard deviation is $\boldsymbol{\sigma}$ (sigma). • Population parameters are used to establish *historical performance*.
Sample Size (n)	• A representative and *random* and *independent* group of CTQ values drawn from a population.
Subgroup	• A group of CTQ values measured during a single process run for use in a Control Chart.
Sample Statistic	• A summary number calculated from a CTQ sample that is used to estimate population parameters. • The sample statistic for average is $\bar{X}$. • The sample statistic for standard deviation is $\boldsymbol{s}$.
Control Chart	• A graphical method that plots summary statistics of CTQ samples plotted in chronological order as the process was run in production. • It is a time-based graph that shows whether a process is stable by comparing its measured variation to statistically calculated control limits.
IC	• The process is in In-Control.
OOC	• The process is Out-Of-Control.
IS	• The production unit(s) is In-Specification.
OOS	• The production unit(s) is Out-Of-Specification.

Term	Definition
Special Cause Rules	• Rules applied to CTQ samples plotted chronologically on a Control Chart to determine if that process remained stable. • A Special Cause Rule "violation" occurs when a point or pattern on a control chart shows **non-random behavior**, meaning the process is being influenced by something *outside* its normal, stable variation. • A rule violation indicates that the process should be considered out-of-control (OOC). • An OOC process indicates that the process has shifted and is no longer delivering its historical performance (μ, σ).
Stability Index (S_i)	• A new index introduced in this book. • Calculated as number of process runs that pass Rule 1 (1 Point out of Control) divided by total number of process runs under consideration. • Count the number of control chart Process Runs that pass Rule 1, divided by the total number of process runs under review.
85% Rule	• A new rule introduced in this book. • If the *Stability Index* (S_i) ≥ .85, then the process considered to be *well-behaved* and *reasonably stable*. • It is calculated by converting the *Stability Index* to a percentage.
Stable Process (In-Control) (IC)	• A process performing with only common causes present, as evidenced by a control chart with no Special Cause Rule violations. • A stable process is considered in-control (IC).
Out of Control Action Plan (OCAP)	• Actions defined by the SPC Chart owner or the organization that are executed whenever an OOC event occurs (violation of a special cause rule) with the intent to determine and eliminate the cause for the OOC event. • Effective OCAPs will return the process to deliver its historical performance and remove the special cause of variation (OOC event).

Term	Definition
Out of Specification Action Plan (OSAP)	• Actions defined by a specification owner or the organization that are executed whenever an OOS event occurs. • Effective OSAPs will contain OOS material and specify disposition actions such as scrap, rework etc. • OSAPs should not include any process adjustments.
Production Unit	• The smallest measure of output or product, typically individual items, which are manufactured or processed.
Production Lot	• Refers to a batch of production units collectively traveling together throughout manufacturing. • Production Lots typically have unique tracking numbers used in the production software and are managed separately in inventory. • Typically, the production units within a production lot share the production lot numbers.
Process Run	• Refers to the single execution of a manufacturing process step. • Process runs may contain production units (single or multiple) or production lots (single, partial, or multiple) during that single execution of the manufacturing process. • Process Runs are those events that should be monitored by SPC Charts.
Reducing Variability	• SPC (Statistical Process Control) charts are used to identify and address special cause variation, leading to a reduction in overall process variability and improvement in process performance once the special cause has been identified and eliminated.
Subgroup	• CTQ data collected for each individual process run.
Variation	• Refers to the differences or changes in a specific characteristic or trait among individuals or data points within a process. • Variation is the observed fluctuation in the data. • Common Cause Variation is natural, inherent variation that is stable and predictable over time. • Special Cause Variation is an unusual or non-routine variation that arises due to specific, identifiable causes, events, or factors.

Term	Definition
Variability	• Refers to the extent to which these differences (or variations) occur. It measures how much the data points in a set differ from each other. • Variability is the underlying characteristic of the process that produces that fluctuation.

Let's now examine the characteristics of a typical Process Control Chart, consider Figure 1.

Figure 1 Typical Control Chart

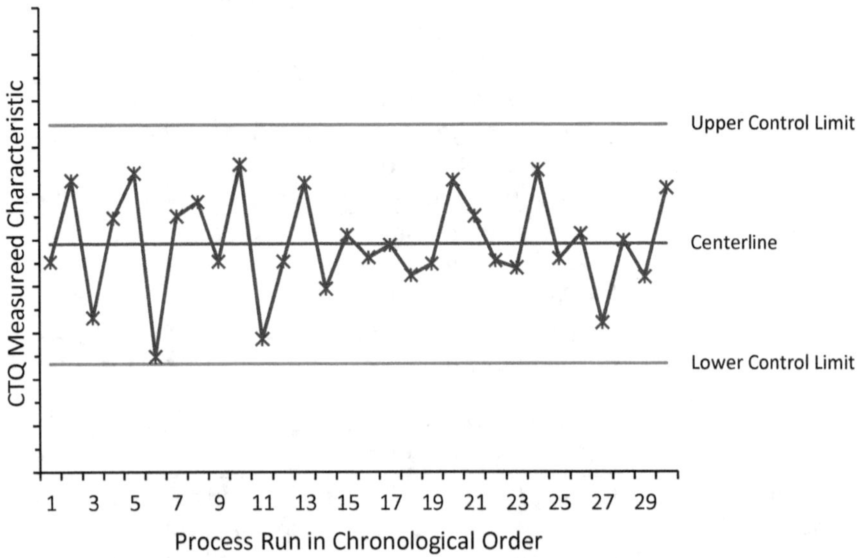

The control chart shows each measured CTQ sample value (X) plotted chronologically for each process run as they occurred in production. The chart contains a *Centerline* that represents the process average (population parameter μ) calculated during the control chart setup. The charts also contain an *Upper Control Limit (UCL)* and a *Lower Control Limit (LCL).* These control limits are also calculated during control chart setup such that almost all the individual sample data points used during control chart setup fall within these limits. Typically control limits are +/-3 times the calculated standard deviation (population parameter σ) determined during control chart setup and are typically referred to as 3-sigma limits.

If the control chart is setup properly, then when a CTQ data point falls outside the control limits, the probability of that happening by chance is very small, so we assume the process is out-of-control (OOC). For example, if the underlying CTQ distribution is a normal distribution the probability of a point being outside the upper or lower control limits (+/-3-sigma) is 0.3%. So, when this happens, we assume a special cause happened and the process is no longer delivering its historical performance. This OOC process warrants an investigation to determine and remove the special cause for the OOC event thereby returning the process to its historical performance. This is done by initiating an Out-Of-Control Action Plan (OCAP).

Notice the added zones in Figure 2. The "zones" represent +/- 1 standard deviation (sigma), +/- 2 sigma or +/- 3 sigma away from the centerline.

Figure 2 Control Chart Zones

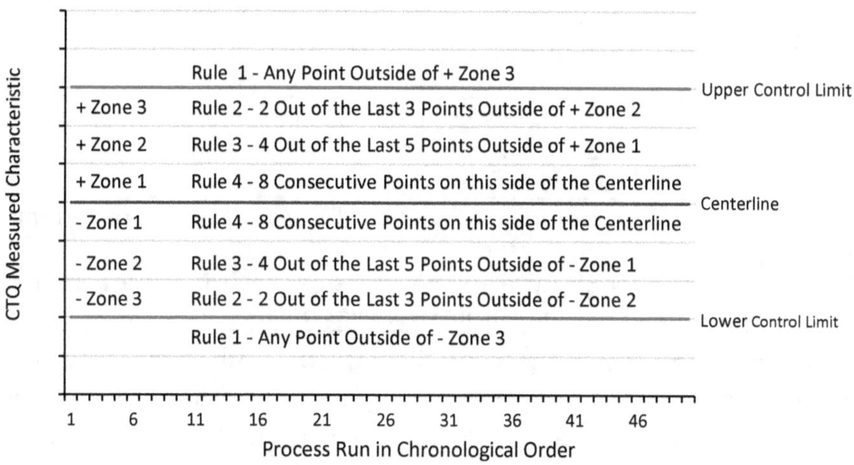

Special Cause Rules (numbered 1 through 4) for an underlying normal distribution are also depicted in Figure 2. Other Special Cause rules can be found in literature, but these are the most common. Notably, other rules may be needed when the underlying distribution is not a normal distribution. The premise is the same, if one of these rules is found true (i.e., a rule violation), the probability of that occurring by chance is so small, we shall assume it did not occur by chance and is the result of a special cause. Whenever a violation occurs, the process is judged to be OOC and warrants an OCAP to return the process to its historical performance. (NOTE: The good news is most production software allows the SPC Chart owner to enable or disable special cause rules.)

Over the years, statisticians have developed various Control Chart types for variable data. Table 2 provides a list of the most common control chart types for variables. Of course, there are many other control chart types. For example, there are specific control charts for dealing with processes with non-normal distributions. These require mathematically transforming data, which makes them more difficult to use and setup. The good news is that I can say *from experience* that when setup properly, all these control chart types work. In addition, most of these chart types are available in modern production SPC software.

Table 2 Common Control Chart Types for Variables

Chart Type	What it Tracks - Other Key Information
X_i & MR	• Individual measurements and the range between two consecutive measurements.
$\overline{\boldsymbol{X}}$ & R	• Subgroup average and range within each subgroup.
$\overline{\boldsymbol{X}}$ & s	• Subgroup average and the standard deviation within each subgroup.
CUSUM	• Cumulative summation if deviations from target. • Effective at detecting small shifts in the subgroup average from a specified target value.
EWMA	• Exponential Weighted Moving Average of current and past data. • Effective at detecting small shifts in the subgroup average from overall population average.

The better news is you only need 3 chart types to have an effective SPC Program for variables: X_i *& MR,* $\overline{\boldsymbol{X}}$ *& R, and* $\overline{\boldsymbol{X}}$ *& s* chart; and, if you follow my setup methods proposed in this book you most likely will not need to transform any data even if the underlying distribution is not normal. Table 3 provides my Control Chart type recommendations. Why is this important? When you are trying to coach an organization to migrate from "SPC for Show" to "SPC for Improvement" it's simpler to train personnel (engineering, quality & production personnel) and deploy throughout a factory when using less chart types - especially when they are known to work!

There is another very important reason why I say you only need these 3 control chart types - even if the underlying distribution appears not normally distributed during control chart setup. From my experience, I can confidently say that these initially non-normal processes quickly became normally distributed when control chart owners followed the methods described in this book.

Table 3 Recommended Chart Types to Use

Chart Type	Recommendations
X_i & MR	• Useful when only individual production units are processed in each run (no rational subgrouping). • Highly recommended for Maverick Lot Programs - See Chapter 13.
$\bar{X}$ & R	• Useful when the control chart subgroup size ≥ 2. • Also useful for other subgroup sizes when operators manually plot control charts without the aid of production SPC software.
$\bar{X}$ & s	• Useful when the control chart subgroup size > 2. • Preferred when production SPC software is available to capture raw sample information and immediately plots the control chart.

All three control chart types are based on the Normal Distribution and include 2 key assumptions:

1. Each sample is randomly selected and independent of its predecessor or successor.
2. The measurement procedure is the same for each sample and taken consistently from sample to sample.

Let's review the key formulas in Table 4,5 and 6 for each recommended control chart type.

Table 4 Key Formulas for Individuals, Moving Range Chart

X_i = Current individual measurement X_{i-1} = Previous Individual Measurement $MR_i = X_i - X_{i-1}$	$\bar{\bar{X}} = \frac{\Sigma\ (X_1 \ldots X_k)}{k}$ $\overline{MR} = \frac{\Sigma \lvert X_1 - X_2 \rvert + \lvert X_2 - X_3 \rvert \ldots . \lvert X_{k-1} - X_k \rvert}{k-1}$ where the difference between the 2 readings is the absolute value and k = number of moving ranges
$\bar{\bar{X}} = Centerline$ $LCL_{\bar{X}} = \bar{\bar{X}} - E_2 * \overline{MR}$ $UCL_{\bar{X}} = \bar{\bar{X}} + E_2 * \overline{MR}$ $UCL_{MR} = D_4 * \overline{MR}$ $LCL_{MR} = D_3 * \overline{MR}$	$\mu = \bar{\bar{X}}$ $\sigma = \frac{\overline{MR}}{d_2}$

Since this control chart type is based on two consecutive individual measurements, the moving range subgroup size is effectively 2, and the constants are always $E_2 = 2.66$, $d_2 = 1.128$, $D_3 = 0$ and $D_4 = 3.268$.

Table 5 Key Formulas for X-Bar, R Charts

$\bar{X}_i = \frac{\sum (X_1 .. X_n)}{n}$, $R_i = X_{Max} - X_{Min}$ where n = number of observations per subgroup	$\bar{\bar{X}} = \frac{\sum (\bar{X}_1 \dots \bar{X}_k)}{k}$, $\bar{R} = \frac{\sum (R_1 \dots R_k)}{k}$ where k = number of subgroups
$\bar{\bar{X}} = Centerline$ $LCL_{\bar{X}} = \bar{\bar{X}} - A_2 * \bar{R}$ $LCL_{\bar{X}} = \bar{\bar{X}} - A_2 * \bar{R}$ $UCL_R = D_4 * \bar{R}$ $LCL_R = D_3 * \bar{R}$	$\mu = \bar{\bar{X}}$ $\sigma = \frac{\bar{R}}{d_2}$
Where A_2, D_3, D_4 and d_2 are taken from Table 35 Constants for X-Bar, R Charts in Appendix 15.1	

Table 6 Key Formulas for X-Bar, s-Charts

$\bar{X}_i = \frac{\sum (X_1 .. X_n)}{n}$, $s_i = \sqrt{\frac{\sum (X_1 - \bar{X})^2}{n-1}}$ where n = number of observations per subgroup	$\bar{\bar{X}} = \frac{\sum (\bar{X}_1 \dots \bar{X}_k)}{k}$, $\bar{s} = \frac{\sum (s_1 \dots s_k)}{k}$ where k = number of subgroups
$\bar{\bar{X}} = Centerline$ $LCL_{\bar{X}} = \bar{\bar{X}} - A_3 * \bar{s}$ $UCL_{\bar{X}} = \bar{\bar{X}} + A_3 * \bar{s}$ $UCL_s = B_4 * \bar{s}$ $LCL_s = B_3 * \bar{s}$	$\mu = \bar{\bar{X}}$ $\sigma = \frac{\bar{s}}{c_4}$
Where A_3, B_3, B_4 and c_4 are taken from Table 36 Constants for X-Bar, s Charts in Appendix 15.2	

Most modern SPC software embeds these formulas and constants in their application. So, they are automatically applied when entering the data.

So, what are the primary causes why SPC Charts become ineffective? It is my experience that the 2 most important interrelated reasons why controls become ineffective are: the ***control chart is setup wrong*** and/or there exists ***competing priorities.*** Often both reasons simultaneously exist.

Often the implementors are unaware of these setup errors since they just follow SPC classroom training, recently learned formulas; or are just implementing software program instructions. SPC owners then turn on all or many of the available Special Cause Rules because they are easily enabled in the software. Incorrectly setup control charts along with many enabled OOC Special Cause Rules trigger many false OOC events. Investigative OCAPs then do not appear to work, frustrate SPC Owners. The "*domino effect*" begins.

The "*domino effect*" refers to a series of changes made by SPC Owners when previous actions fail, resulting in ineffective SPC Control Charts. For Example, SPC Owners begin changing OCAPs from investigative actions to "call Engineering" or just "make adjustments." Even worse competing priorities such as the need to "meet-the-spec" cause other SPC Owners to begin putting Specification Limits on Control Charts. The OCAPs start becoming reactions to Out-Of-Spec (OOS) events rather than OOC events. Bottom line - the SPC Charts become ineffective! SPC Chart Owners leave them in place because they are required by the organization or key customers, and ultimately your SPC program becomes "SPC for Show."

In the next couple of chapters, I will discuss each of these "*domino effects*" in more detail. I will also provide *proven* solutions that, when implemented, will ensure *The SPC Promise*: drive process stability and improve process variability by identifying and eliminating assignable causes.

4. COMPETING PRIORITIES: PRODUCT OR PROCESS CONTROL

Prior to implementing SPC Programs, the organization's priority *seemed* simple: Engineering and Production were responsible for developing, implementing, and qualifying processes that delivered product that was in specification. SPC came along and promised that by stabilizing processes, they would deliver more products in spec, improve production yield, lower costs, etc. Customers came along (notably customers in the Automotive and Aerospace industries) and required customers to meet specifications and to implement SPC programs to become a qualified supplier. The stage was set for an organization to attempt to implement effective SPC programs, but it also set up competing priorities between product control and process control. It turns out that this competing priority is ***one of the main reasons*** why SPC Programs (and their Charts) become ineffective.

Let's examine these competing priorities, consider Table 7.

Table 7 Competing Priorities

Product Control	**Vs.**	**Process Control**
• Manufacture product CTQ within specification	Primary Objective	• Continuously minimize CTQ variation around a target value
• Acceptance Sampling Methods (AQL)	Statistical Tools	• SPC Control Charts
• Detect and contain non-conforming (OOS) production units	Primary Strategy	• Achieve and Maintain Stable processes • Enable Ship-to-Control
• Consistently deliver outgoing product quality to the AQL level	Desired Outcome	• Prevent production of non-conforming (OOS) production units

In the world of Product Control, the organization is supposed to manufacture products and only deliver products within specifications. Statisticians developed methods such as Acceptable Quality Levels (AQL) to assist.

The Acceptable Quality Limit (AQL) is a statistical measure used to determine the maximum number of defective items allowed in a sample batch before the entire batch is rejected. AQL standards provide predefined sampling plans that specify the sample size and the acceptable number of defects for different inspection levels. AQL standards classify defects into categories such as critical, major, and minor. Each category has its own acceptable limit, ensuring that only batches with an acceptable level of defects are accepted. AQL provides clear criteria for accepting or rejecting a batch, simplifying the decision-making process. If the number of defects in the sample exceeds the acceptable limit, the batch is rejected.

Samples are drawn from a population using the AQL criteria, and the organization contains any non-conforming (OOS) material per the acceptance criteria. This approach maintains the quality of shipped material to the selected AQL level. However, some customers started noticing that in spec (IS) products that were also OOC often had early life reliability failures or other latent quality issues. Their countermeasure to this situation emerged as a "Ship-to-Control" requirement. "Ship-to-Control" is a customer-driven requirement where a supplier must demonstrate tight process control and defect prevention before being allowed to ship product without additional customer inspection. Not only do customers want their products built to spec, but they also want the process runs to be in-control.

In the Process Control world, the goal is to continuously minimize CTQ variability around a target value. Statisticians developed methods such Statistical Process Control (SPC) Charts to assist. The primary strategy is to prevent the production of non-conforming (OOS) production units by stabilizing processes. Well-defined specifications (see Chapter 11. Calculating Product Specifications) in combination with stable processes allows the organization to implement the "Ship-to-Control" requirements and deliver In Specification product. These approaches promise to continuously improve product quality.

Let's further examine how these priorities *are supposed* to work together. Product CTQs should have specification limits and CTQs monitored by a control chart should have control limits. Table 8 compares Control Limits and Specification Limits. It describes how the limits should be derived and what they are supposed to be used for. (Note the Specification Limits comments apply to both double sided and single sided product CTQs.)

Table 8 Control Limits vs Specification Limits

Control Limits (UCL, LCL)	Specification Limits (USL, LSL)
• Calculated from actual historical process performance.	• Specified during the product design phase based on required product performance.
• Applies to sample statistics (subgroup) drawn from process performance CTQs which are plotted on a control chart.	• Applies to all individual production units.
• Used to determine if the process is stable, i.e., continues to deliver its historical performance.	• Not used to make decisions on process performance and/or stability.
• Used to determine if an OCAPs should be run. • Not used to determine if an OSAP should be run.	• Used to determine if an OSAP should be run. • Not used to determine if OCAP should be run.
• Used to determine if production units can be shipped when there is a Ship-to-Control requirement.	• Used to determine if production units meet product performance requirements and can be shipped.

Consider the typical SPC situation where you already setup the control chart. You are running a process that delivers the CTQ on production units that have product specifications, and it is also a critical *process* performance characteristic. You run the process and measure the designated samples. Now you are at a decision point on what to do next. Figure 3 depicts how the manufacturing actions *are supposed to work* in the world of competing priorities.

Figure 3 Specification Limits vs Control Limits

ACTIONS

CONTROL			
	IN	$	• Contain Out-of-Spec Material (OSAP) • Keep Running the Process
	OUT	• Ship Material in Spec • Run OCAP	• Contain Out-of-Spec Material (OSAP) • Run OCAP
		IN	OUT

SPECIFICATIONS

The Y-Axis designates whether the process is In-Control (IC) or Out-Of-Control (OOC). The X-Axis depicts whether and samples were found to be In-Spec (IS) or Out-Of-Spec (OOS). The boxes indicate what is supposed to happen for each of the 4 different cases.

Case 1: The process run is IC and production units are IS. This is the optimum situation, no action needed, continue processing, make some $!

Case 2: The process run is OOC, but production units are IS. You should ship the material in spec and run an OCAP to bring the process back in control. (Except for customers that have a "Ship-to-Control" requirement.)

Case 3: The process run is IC, but some production units are OOS. You should contain the OOS material using an OSAP (Out-Of-Specification Action Plan) and keep running the process (OCAP not needed) without process adjustments. NOTE: Most organizations often find it easy to contain OOS material but find it very difficult to let the process keep running if it is making OOS material.

The worst case is of course Case 4: The process run is OOC, and some production units are OOS. You should contain the OOS material using the OSAP and run the OCAP to bring the process back in control.

For each of the OOC cases where the situation calls for an OCAP, the OCAP is supposed to stop the process, investigate, and remove the special cause for the OOC event, then resume processing. Because of this competing priority (product versus process control), organizations often want to do something more than just containment when they detect OOS material. They often integrate these additional OOS actions into the OCAP, often at the expense of determining the special cause. These other actions often render the SPC Chart ineffective. In Chapter 5, I discuss in detail various types of Good, Bad and Ugly OCAPs!

There are some other key issues to consider. Is your specification any good? Can your process meet your current specifications now? Is it likely to meet it in the future? If your process has no hope of meeting your specifications, then making OOS adjustments within an OCAP will only make matters worse (I say this from experience). In Chapter 11, I address these issues relating to specifications. I provide a method to analyze your current process and based upon your current process performance, suggest how to calculate a specification. I also provide a method, once you have released a *qualified* and *well-behaved* control chart to production, to determine what your specification could be once the process achieved stability.

Why would calculating (or re-calculating) specification limits from control charts be a good idea when specifications are usually set during product development and before production begins? Because, from my experience, 80% of the time Product Engineering accepted specification adjustments because they would not affect product performance. They usually measured production samples at the tails of the suggested specifications and confirmed no impact before they approved the specification change. In addition, they are also useful in establishing specifications for future products that use the same process.

There are two more topics I want to discuss that I found helps resolve this competing priority. Many CTQs characteristics serve as both the product and process characteristic.

Consider the following process: a chemical vapor deposition (CVD) process deposits a very specific composition and thickness of silicon nitride on a substrate material, such as a silicon wafer. There is a LSL and a USL for the deposited Nitride film thickness. After the process is run, thickness measurements are taken from samples and plotted on a control chart. Are the samples IS or OOS? Is the process IC or OOC? The OCAP versus OSAP competition begins.

But what is the *process* supposed to deliver? The process is supposed to deliver a uniform Nitride Deposition rate on substrates. You want to control the deposition rate with minimum variation across the substrates. If you have a stable process, you then select the process time to achieve the specified Nitride Thickness. Suppose you plot the deposition rate as the CTQ for this process on a control chart rather than the Nitride Thickness. The control limits are derived from deposition rate data, not final thickness data. Now, there is no competition with the Product Nitride Thickness specification. If the deposition rate goes OOC you run an OCAP that only deals with the process. If some samples are found to be OOS, you run an OSAP that only deals with the disposition of the OOS material. The decisions for OOC and OOS are independent of each other! You can now define effective OCAPs.

Here is another example: The process is a wet etch process where production units are placed in a buffered Hydrofluoric Acid etch batch. The intent of the process is to etch away a certain amount of oxide i.e., deliver a consistent oxide etch rate with minimum variation across the production units. The process engineer ran the control chart setup procedure and released a *well-behaved* control chart for etch rate per procedures described in this book. A couple of months later, he decided to invent an etch rate specification and added the specification limits to the control chart. Apparently the OCAPs were not working. Now the focus has completely changed from maintaining process stability to meeting the etch rate specification. It turned out his original OCAPs were adjustment OCAPs rather than investigative OCAPs, so he was routinely changing the process as he never found the real *special cause*. Even though his actions had good intentions, it caused higher etch rate variation than what was measured during control chart setup i.e., the process variability degraded because of his OCAPs. Once we removed the specification limits and found the special cause for the OOC events the etch rate variation recovered to its performance found during control chart setup. We also rewrote the OCAPs. Six months later the etch rate variability reduced by another 30% due to the effective OCAPs! Bottom line -- there was no reason to have an etch rate spec limit and the type of OCAPs made a *huge* difference. In the next chapter, I will discuss The Good, The Bad and The Ugly OCAPs.

RECAP

From these examples, I found 4 key strategies to help minimize the Product versus Process control competition:

1. Whenever possible, find and separate Process CTQs from Product CTQs.
2. Define separate OCAPs for the Process CTQ and OSAPs for the Product CTQ. Modern production SPC software can handle this separation with ease.
3. Do not invent process specifications when you do not need them. Maintain focus on keeping the process in control. Inventing unneeded specifications will revive the competing priority of product versus control and start the *"domino effect."*
4. If you do need a control chart to monitor a Product CTQ, be sure that OCAPs address process controls and OSAPs are only used to contain and disposition OOS material.

In the real-world meeting product specifications often compete with maintaining process controls. Even if the control chart was set up correctly, this competing priority often adds specification limits on control charts, changes the nature of OCAPs thereby rendering the SPC Chart ineffective. As suggested, the best way to address this competing priority is to separate Process CTQ's OCAPs from Product CTQ's OSAPs, i.e., ensure independent OCAPs and OSAPs. Finally, do not invent unneeded process specifications for Process CTQs.

5. TOP REASONS WHY CONTROL CHARTS ARE SETUP WRONG

As mentioned in the introduction, the vast majority of SPC Charts, when I first review them, were not effective at maintaining process stability. Never mind improving process variability by removing the reason(s) for assignable causes. Often ineffective SPC charts are widespread across an organization yet remain in place to show customers and/or auditors. Because they are not effective, alternative organizational policies begin emerging that make matters worse. As a Quality Executive, I was often responsible for turning "SPC for Show" programs into *effective* "SPC for Improvement" programs. As my quality science skills grew, I began finding solutions for these situations. I must admit I was more successful with some organizations than others. For those individual SPC Chart Owners that adopted these methods, they had the most stable process(es) in the factory that was constantly improving. For those organizations that adopted the methods in this book and deployed them *across* the factories, they experienced widespread process improvements, reduced production costs, and without a doubt business expanded!

In this chapter, I am documenting the Top Five reasons why SPC Charts are setup wrong. In Chapter 7, I provide a detailed method for setting up and "qualifying" a control chart for production which addresses all these setup errors.

From my experience, the Top Five Reasons why control charts are setup wrong are:

1. Control Charts Monitor Production Lots, Not Process Runs.
2. Specification Limits for Individuals are Placed on Control Charts for Averages.
3. The Control Limits are Setup Using the Wrong Sample Size.
4. Evidence of a Lurking Variable that Appears During Control Chart Setup is Not Addressed before releasing the control chart to production.
5. The Good, The Bad and The Ugly OCAPs.

These reasons are interconnected, often causing a "*domino effect.*" The "*domino effect*" is a metaphor that describes a situation where one event sets off a chain reaction, leading to a series of related events or consequences. The term comes from the game of dominoes, where knocking over one tile causes a line of adjacent tiles to fall in sequence. The concept highlights the importance of understanding the interconnectedness of events and the potential for small actions to have far-reaching impacts.

Let's explore each one!

5.1. Control Charts Monitor Production Lots Not Process Runs

Control charts often plot chronological production lots versus process runs on the x-axis. If a process run consists of a single production lot, or a single production unit, no harm is done. If the chart was setup correctly, the control chart would work just fine. However, it gets a little messy if a process run is not a single production lot.

Consider Table 9. Column 1 lists examples of different relationships between a process run and a production lot or unit. Column 2 describes examples of how control chart data is often derived from Column 1 examples. Column 3 describes the potential errors if production lot or unit data is incorrectly plotted on the process control chart.

Table 9 Product Lot vs Process Run

Process Run Consists Of	Control Chart X-Axis Plot	Potential Errors
Single Production Lot	• One process run data point for each production lot.	• None - should work fine.
Single Production Unit	• One process run data point for each production unit.	• None - should work fine.
Multiple Production Units	• All production unit data from a production lot are combined from multiple process runs for a single control chart entry. • One data point for each production lot includes multiple process runs.	• Missing process run data. • Too few data points since not all process runs are plotted. • Causes OOC event errors! • May miss OOC events for a process run.

Process Run Consists Of	**Control Chart X-Axis Plot**	**Potential Errors**
Multiple Production Lots	• Often the production lot data are separated even though they are derived from a single process run. • Each production lot is plotted on the control chart separately.	• Multiple data points for a single process run. • Causes OOC event errors!
Partial Production Lot(s)	• Often the production lot data from multiple process runs are combined. • One data point for each production lot derived from multiple process runs.	• Missing process run data. • Too few data points since not all process runs are plotted. • Causes OOC event errors!

Common theme: there is a high risk of OOC event errors if care is not taken to ensure the data plotted on a control chart represents process runs and not production lots.

These errors became more prevalent as software allowed specification limits on control charts and the emphasis begins shifting from process control to product control.

5.2. Specification Limits for Individuals are Placed on Control Charts for Averages

Often I saw control charts like the X-Bar chart in Figure 4!

Figure 4 X-Bar Chart with Spec Limits

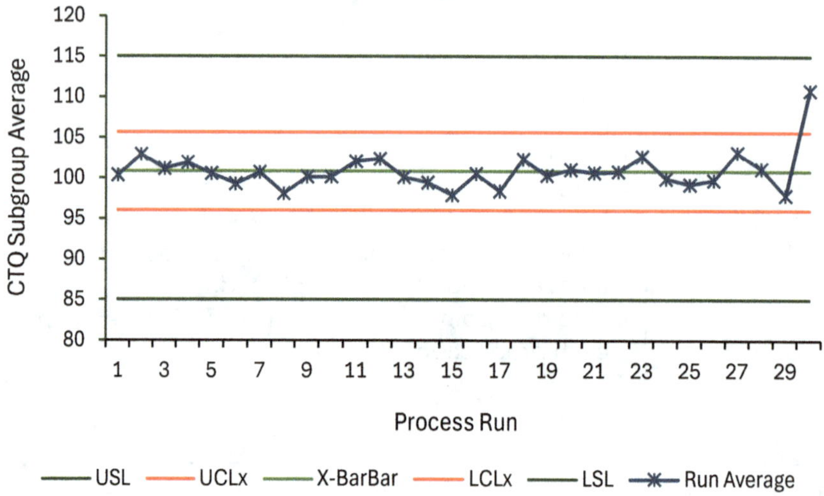

Notice this chart includes an *Upper Specification Limit (USL)* and a *Lower Specification Limits (LSL)*. It does not matter whether you have double sided specs (USL & LSL) or single sided specs (USL or LSL), the same principles/analysis applies.

I started seeing these types of charts when the software allowed it, and the CTQ is an actual product characteristic which had specification limits. The problem is that control charts, especially X-Bar charts, are for averages, whereas product characteristics specifications are for individual production units. In my career, I do not remember any product CTQs that had specifications set for averages.

So why are they putting spec limits for individuals on control charts for averages? I found 3 simple reasons. First, the organization only wanted to run OCAPs if any of the measured samples within the subgroup were out-of-spec because meeting specifications were judged as more important than maintaining process stability (See Chapter 4 on Competing Priorities). Second, customer requirements and auditors, notably from the Automotive and/or Aerospace industries often asked and sometimes demanded that they be placed on Control Charts for certain Product CTQs. Finally, modern SPC software allows showing USL/LSL for any CTQ characteristic that has spec limits - just turn it on!

I actually approached several SPC software providers and asked if they were aware of the contradiction (UCL_x , LCL_x are for averages and USL and LSL are for individuals). They indicated that they were aware of this contradiction but added the feature because of customer demand.

In Figure 4 X-Bar Chart with Spec Limits we see that run #30 is OOC since it fails Rule 1. This OOC event should cause an OCAP. The type of OCAP is critical. If the OCAP investigates and removes the special cause - then all is good, the process performance will be restored - and potentially improved. However, what I found was that when spec limits were placed on control charts, the nature of OCAPs changed. My audits of work instructions in cases like this often-revealed OCAP work instructions directed operators to examine the samples from the run. If none were found to be out-of-specification, then ignore the OOC event and continue processing! If a sample was found out-of-spec they were more often told to execute an adjustment rather than stop the process and figure out the special cause. The OCAP instructions often included instructions for the operator to scrap or rework the out-of-spec production unit(s).

So why is it ignoring the OOC event if no samples are found out-of-spec such a bad idea? A core premise of control charts is that the process continues to deliver μ and σ (runs #1 through 29) until you have evidence that it is not, i.e., the OOC event (run #30) provides that evidence. During control chart setup, estimate the population parameters μ and σ (see Table 4 for Individuals & Moving Range Chart, Table 5 for X-Bar & R Charts, or Table 6 Key for X-Bar & s-Charts for key formulas). So, since run #30 is out of control, the process is now delivering a different distribution. So, what is the best estimate of the new distribution? It is μ_{30} and σ_{30}! Let's explore this example.

For the process represented by Figure 4 X-Bar Chart with Spec Limits, suppose the USL = 115, LSL = 85 with the center of Spec = 100 as indicated on the chart. During control chart setup, you determined μ ($\bar{\bar{X}}$)= 100.7 and σ ($\frac{\bar{s}}{c_4}$)= 5.16 from the 30 measured subgroups (in this case I modeled 10 samples per subgroup). I plot the theoretical normal cumulative distribution of individuals in Figure 5.

Using Z-Scores, I calculated the percentage out-of-spec for the stable process to be .4%, mostly on the high side.

Figure 5 Distribution of Individuals (X-Bar = 100.7 and Std = 5.2)

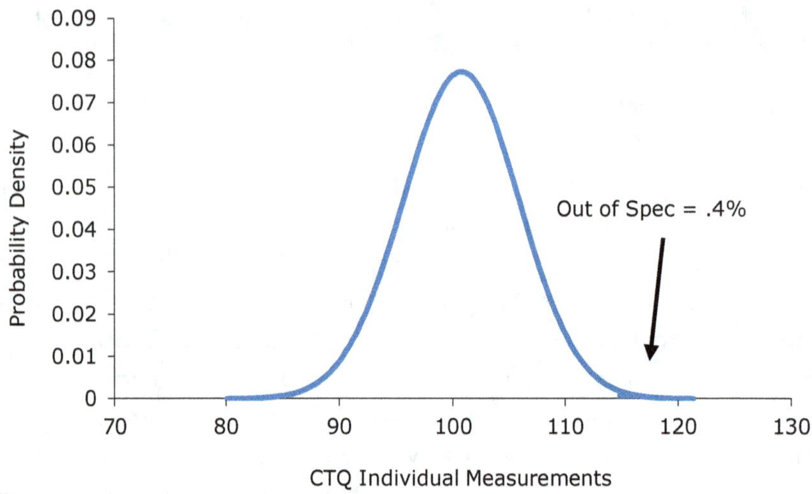

Now consider the OOC run #30. What is the best estimate of the new distribution? Your best estimate is to take the OOC event samples and calculate μ_{30} and σ_{30}. In this case that turns out to be $\mu_{30}\,(\bar{\bar{X}})$ = 110.8 and σ_{30} $(\frac{\bar{s}}{c_4})$= 4.4. Figure 6 depicts the theoretical cumulative distribution of individuals for the OOC population.

Figure 6 Distribution of Individuals (X-bar = 110.8 and Std = 4.4)

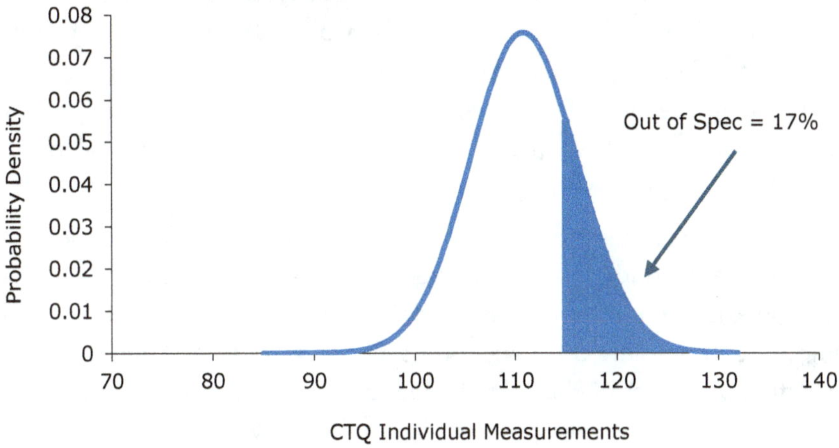

The chart indicates you are now delivering 17% out-of-spec, again on the high side. So, what are the ramifications of the OCAP instructions? If the instructions were to let the process continue if none of the individual data points from run #30 were out-of-spec, you are now building >17% of your product out-of-spec. This is despite subgroup average the chart showing the process "in spec" and OOC.

Bottom Line: Specification Limits for individuals do not belong on control charts for averages! They drive the wrong OCAP behavior.

5.3. The Control Limits are Setup Using the Wrong Sample Size

It seems so simple, you take samples from a process run, measure the CTQ characteristic, then follow the control chart setup rules for that chart type, sample size, and bingo - you have a useful Control Chart! The Control Chart formulas adjust the control limits for sample size. Even better, modern SPC software allows you to just plug in your numbers and make the calculations for you. Seems reasonable - except most control charts (certainly those recommended in Table 3) follow principles of the Central Limit Theorem. There is a very important assumption that often gets overlooked during control chart setup. Recall a key assumption, *samples drawn from the process are assumed to be random and independent.* Most engineers interpret that by just picking what they judge as random samples across a process run will satisfy this assumption. There is also an inherent belief by process owners that more samples are better. However, my experience shows, if the samples taken violate this assumption, not all sample data is *effective* for purposes of a control chart. You start with wrong control limits because the control limits are not based on *random* and *independent* samples. Wrong limits cause false OOC events! The "*domino effect*" begins! Let me show you what I mean.

Here is the process: Each process run consists of 50 production units which is one production lot. You take 18 samples per process run to make sure you know what the process is doing and measure the CTQ for each sample. You complete 30 process runs, the recommended number of runs for setting up a control chart. At 18 measurements for 30 runs (18 samples x 30 subgroups = 540 datapoints) you have plenty of data to setup your control chart. Your software gives you the following $\bar{X}$ & s charts depicted in Figure 7. Since the normal distribution is a 2-parameter distribution, you get a chart for assessing the process average and a second chart assessing the process variation (standard deviation) i.e., an $\bar{X}$ *& s* chart.

Figure 7

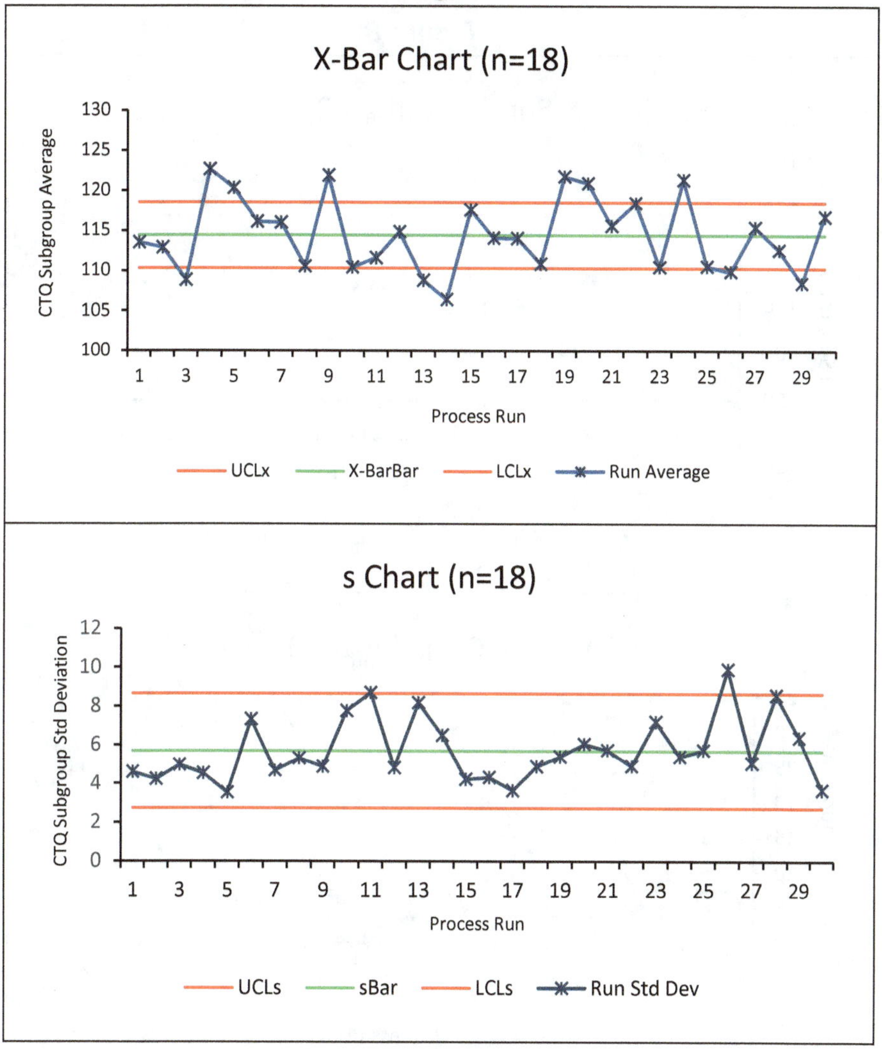

Upon initial review, it looks pretty good - except 13/30 process runs fail Rule 1 (1 point outside of Zone 3, either higher or lower), never mind the other rules, or 43% of your process runs already appear to be OOC. This means you are totally OOC and will need to run a lot of OCAPs!

However, what I learned is that it is more likely that the samples taken are not *random* and *independent* which caused the control limits to be wrong. I figured out a technique to verify the *effectiveness* of samples taken during control chart setup and called it the *Effective Sample Size* (n_{eff}). Once I analyzed samples and adjusted the actual subgroups for the *effective sample size* (n_{eff}), I got the chart in Figure 8.

Figure 8

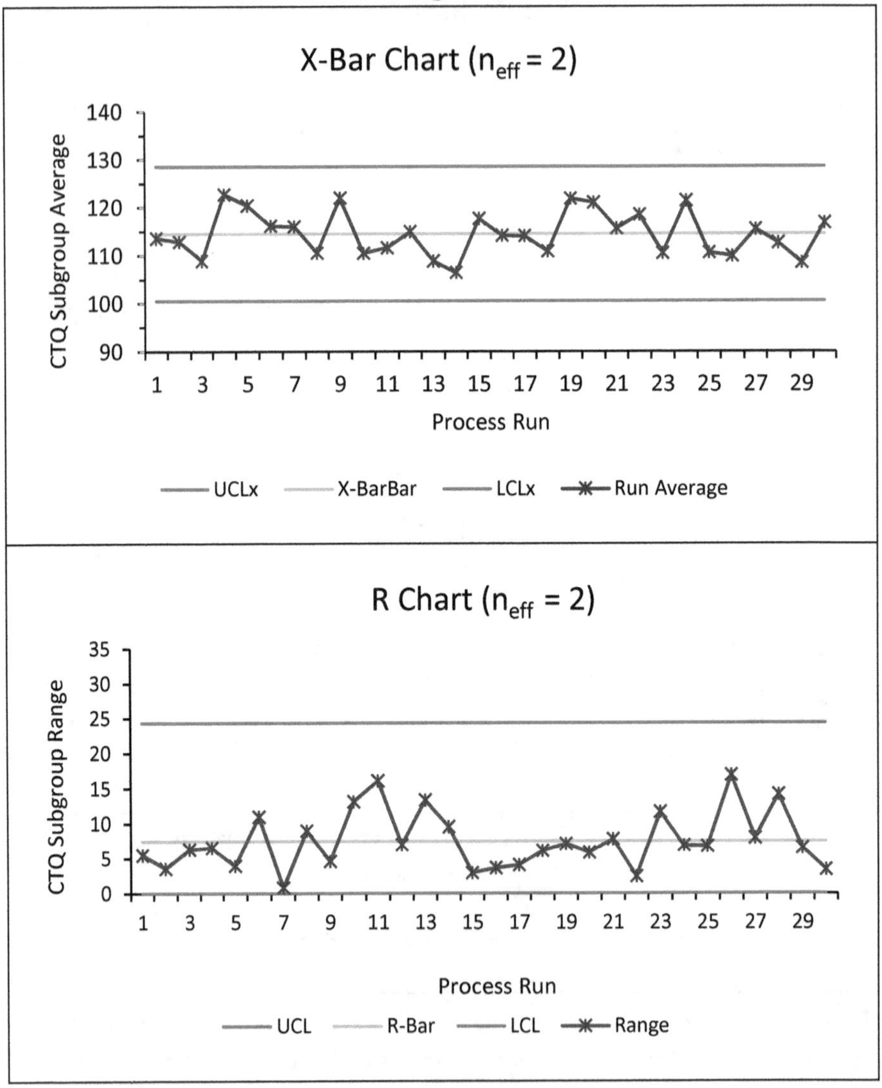

This is the **same data** that was plotted in Figure 7 except with an *effective sample size* (n_{eff}) of 2. Notice all points are in control - no OCAPs are needed! The control chart appears "*well-behaved*" and the process is stable. This is a huge difference from Figure 7 - especially since it used the same data!

It is my experience that using the wrong sample size in your calculations is the #1 *root cause* for why control charts are setup wrong. When this happens the "*domino effect*" begins! This is so important that I devote 2 chapters to this issue. In Chapter 6, I review that exact case on how I discovered (n_{eff}), and why it works. In Chapter 7, I put it all together and provide a detailed method on how to Qualify a control chart before releasing it to production by analyzing the setup data for the *effective sample size* and other factors.

These false OOC events resulting from control chart setup errors cause the organization to either run too many OCAPs, ignore OCAPs or invent and adopt OCAP policies that further makes control charts ineffective. These actions and policies, although well intentioned, usually cause the problem to get worse. When one policy does not work, another policy is introduced, causing the *"domino effect."* The balance of my list includes many of these policies, often defined during control chart setup, that also cause SPC Charts to become or remain ineffective.

5.4. Evidence of a Lurking Variable is not Addressed During Setup

Again, during control chart setup, and before you release a chart to production, you should examine the chart to determine if there are any obvious *Lurking Variables*. An obvious *lurking variable* is evidence of a currently unknown assignable cause that is observed during control chart setup that is impacting the overall process distribution. Figure 9 shows an example of an $\bar{X}$ & s with lurking variable(s) detected during control chart setup.

Figure 9 Lurking Variable Example

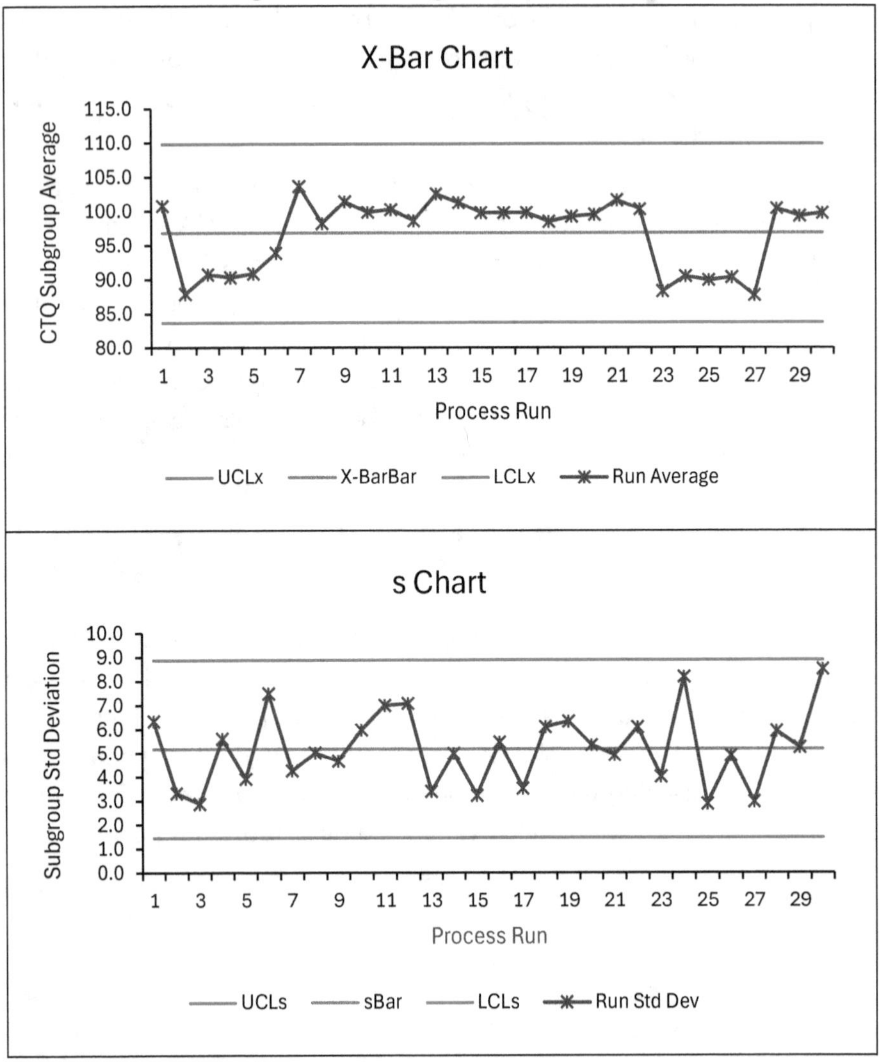

Lurking Variables during control chart setup can be found by applying Rules 2, 3 and 4 to the control chart data during setup. The example depicted in Figure 9 indicates runs #2-6 failed Rule 3, runs #7-22 fail Rule 4 and runs #23-27 fail Rule 2. That is 3 different OOC events indicating the possibility of up to 3 different *lurking variables*! The good news is my experience indicates that OOC events evident during control chart setup are most likely caused by the same *lurking variable*. These *lurking variables* should be addressed before this control chart is released to production. Most of the time, the lurking variable appears on the X-Bar chart rather than the s-Chart. When this happens, the control chart should not be released to production until the lurking variable is identified and addressed.

> A ***Lurking Variable*** in control charts is a main effect factor that causes a noticeable shift to the CTQ above and beyond common causes of variation (a special cause). *Lurking Variables* are identified *during control chart setup* when setup subgroup run data fail Rules 2, 3 or 4.

From my experience, here are the most common *lurking* variable special causes found during control chart setup:

1. *Measurement Error:* Ensure that the CTQ measurement equipment(s) is properly calibrated and/or Gage R&R (Repeatability & Reproducibility) studies have been completed such that you are confident that all measurement equipment and personnel provide consistent and accurate data. Suggestions: (1) only use a single calibrated measurement device for the initial control chart setup; or (2) conduct Gage R&R studies before control chart setup on all measurement equipment that will be used for the process assessment.
2. *Control Chart Plots Production Lots Not Process Runs.* When a process run consists of multiple production lots, and if the x-axis is production lot instead of process run, then multiple data points may be plotted for the same process run which often appears as a lurking variable.
3. *Inconsistent Process Execution:* Sometimes different production personnel and/or process engineers run the process differently. This often happens on different shifts when work instructions and/or training are inadequate.
4. *Inconsistent Process Adjustments:* Often work instructions allow for process adjustments based on measurements taken. If inconsistent process adjustments are applied, the effect often appears as a lurking variable during control chart setup.
5. *Equivalent Process Equipment plotted are plotted on the same chart too early:* This is the most frequent special cause for a *lurking variable* during control chart setup, especially if the possibility of measurement error has been eliminated. Suggestion: for the initial control chart setup, select the primary processing equipment for the initial 30 runs. Use the methods described in Chapter 7 to determine if data from equivalent processing equipment can be combined.

Most of the time in production, there are multiple pieces of equivalent equipment that are supposed to deliver the same process. Some equivalent equipment are "Copy Exact" and others are "Copy Similar." "Copy Exact" means *identical equipment* are setup to run all process parameters *exactly* the same. "Copy Exact" equipment has the best chance of being combined on the same control chart. "Copy Similar" are equivalent equipment, either the same or different from the primary, where the process parameters are equipment specific with the intention of delivering the same process. You can often find out by reviewing the work instructions to see if process parameters are specified by equipment identification. If you find operating parameters are specified by equipment in work instructions, at best you are "Copy Similar." Another indicator is if you find *different* work instructions for equivalent equipment, again in the best case you are "Copy Similar." "Copy Similar" equipment rarely passes tests to be immediately combined on a control chart with other equipment. Review Chapter 8 on useful techniques for combining equivalent equipment.

Although it is very desirable to have all equipment delivering the equivalent process on the same control chart, be careful. If you follow the guidelines in Chapter 7, you will automatically analyze this phenomenon during control chart setup. If different equipment delivers different distributions, put each piece of equipment on its own control chart until the population parameters for each piece of equipment are found to be equivalent. The good news is that modern SPC Software easily allows separate charts for separate equipment.

Bottom line: make sure all equivalent equipment *actually delivers* the same process performance *before* you combine them on control charts.

5.5. The Good, The Bad and The Ugly OCAPs

As previously discussed, the best Out-of-Control Action Plan (OCAP) is a quick response that investigates the reason for the OOC event, removes the cause, then resumes the process. If the special cause was hidden within the previous distribution by virtue of chance, locating and removing the special cause will remove that component of variation, which reduces the overall distribution - thereby improving the variability of the process!

Let me describe 2 real examples of OCAPs triggered by OOC events: Case 1 - The Task Force, and Case 2 - The Process Adjustment. These cases demonstrated the effect of the different OCAP strategies.

CASE 1 – THE TASK FORCE

I was working for a semiconductor company as a Consulting Engineer. We were manufacturing and releasing a new microprocessor in our wafer fabrication factory. Each production lot for the microprocessor took about 10 weeks to complete as it had over 100 processing steps. So, it took a while to see the effects of any process changes. The final test yield was running a very poor 30%, making the microprocessor uncompetitive in the marketplace. The organization asked me to head up a Task Force to study the causes of the poor yields with the intent to significantly improve the yield. The first action we took was to conduct a correlation study of final test yield with process steps. From this correlation study, we found a processing step that seemed to have a high correlation with final test yield. This process did not yet have an SPC Chart deployed. (Side Note: the organization was just beginning SPC deployment throughout the wafer fab.) So, we started the control chart setup work. We collected sample data for 30 process runs, analyzed the data for effective sample size. Once satisfied, we released the chart to production.

This was a furnace process, which consisted of 3 heating zones and could accommodate 3 production lots at a time. One lot consisted of 50 wafers where each wafer included over 100 microprocessor computer chips. So, each process run affected a lot of microprocessors (15,000)! The process grew "oxide" on the wafers, a critical product characteristic. When we saw an OOC event, we stopped the process to investigate the cause, fixed the problem, then resumed production. Two special causes quickly emerged. Cause 1 was an operator who would not wait for the furnace to reach temperature before he put the wafers in the furnace and ran the process. This typically happened on the first shift, when the furnace was first turned on (it was a 2-shift operation). He apparently was being pushed by production management to get the product moving so he would not wait for the furnace to heat up to temperature. The oxide growth rate was slower while the furnace was heating up. We spoke to his management and got agreement that he would wait until the furnace reached temperature and put a buzzer on the furnace to announce when it was safe to run the process. Cause 2 was an intermittent flow meter which provided gases to the furnace - which also affected the oxide growth rate. We replaced the flowmeter and put a control chart (individual's chart) on the flowmeter's output to detect flowmeter OOCs events. We resumed production and monitored the control chart daily.

We continued data analysis across the wafer fab looking for other possible processes affecting final test yield, while we waited for new product to be processed through the furnace and finished through final test. We continued to deploy *well-behaved* control charts on other processes. Of course, it took weeks for the newly processed material to reach the final test. As the newly processed lots began hitting final test, the yield started jumping - within weeks of seeing new production lots from the in-control furnace process we were consistently getting >80% production yields.

As a Task Force, we were really happy - the OCAPs worked! It turns out the company was even happier. They were so happy they sponsored a 3 day get-way to New York City where we enjoyed multiple Broadway Shows for all Task Force members! By the end of the year, the final test yields reached >90% and they asked us to finish deploying SPC charts throughout the fab. Never had a problem with OCAPs that stopped the process and investigated the special cause for an OOC Event.

CASE 2 – THE PROCESS ADJUSTMENT

For this case, I was the Quality Executive for this company that made LED Chips. We were well into our SPC deployment efforts. I had already delivered many SPC training classes and taught many of the techniques contained in this book. Our biggest obstacle was coaching process engineering on defining effective OCAPs.

In this case our product yield was running just fine - then a yield bust hit. We went from consistently getting >90% yield to intermittent lots barely hitting 40%. Our Yield Enhancement Team was on it and began their investigation. They correlated the final test yield to one piece of equipment that deposited, using a chemical vapor deposition (CVD) process, a thin dielectric layer on the wafers. We had multiple pieces of equipment that delivered the same dielectric, yet correlation analysis linked it to a single piece of equipment. We looked at the control chart, which monitored the dielectric layer thickness, and it appeared in control, which was confusing. Then one of the Yield Engineers noticed an OOC event occurred at the beginning of the yield bust. So, he researched the OOC event and the OCAP. Turns out the OCAP instructions just said to "Call Engineering." The on-duty

Process Engineer was responsible for investigating and defining a just-in-time OCAP.

In this case, due to production pressure, rather than stopping the process and investigating the special cause, the Process Engineer knew the problem was the dielectric thickness, so he authorized an untested process change that would bring the dielectric thickness back in spec. He never investigated if the process change would interact with any other subsequent process steps. It turned out that the process change brought the thickness back into spec but changed the dielectric composition and negatively interacted with subsequent processing steps. So, we immediately took this piece of equipment off-line. Final test yields started recovering as this equipment was off-line and the pipeline was filled with product from the other equipment. Eventually, the special cause was discovered, and the equipment was brought back on-line. So, we lived through the yield bust and recovered from an OCAP that allowed an unproven process change.

The point of these 2 cases is to highlight the importance of Out-Of-Control Action Plans. So even if you follow the methods in this book for setting up your control chart, OCAPs make a huge difference on whether SPC control charts are effective.

There are Good, Bad and Ugly OCAPs. Of course, the best OCAP is when an OOC event occurs, the OCAP provides quick response instructions to investigate, locate and remove the reason for the OOC event. Production is often not a fan for this type of OCAP, especially if it requires stopping the process when none of the samples appear out-of-spec. Table 10 lists some common OCAP strategies that I observed once an OOC event is triggered and more *importantly the observed effects* of those OCAPs.

Table 10 OCAP Strategies

OCAP Strategy	**Comments**	**Observed OCAP Effects**
Stop the process, investigate, and remove the reason.	• *Good - Really the BEST!* • Most effective	• Processes quickly achieve stability. • Process variability consistently improves. • Frequency of OOC events consistently reduced.

OCAP Strategy	**Comments**	**Observed OCAP Effects**
All shifts/personnel run the same adjustment OCAP.	• Good • Okay if goal is to maintain current process variability.	• Adjustments will be continuously needed. • Consistent OCAPs sometimes lead to finding the actual special cause. • Process variability improved only if special cause was found. • Frequency of OOC events is reduced if the special cause eventually found and removed. • Since the OCAP adjustment is the same for all shifts/personnel, you very quickly find out if the adjustment is effective. • *If special cause not found, variability remained the same and the frequency of OOC events occurred at the same rate.*
Authorize a *known* process adjustment.	• Good • Okay if goal is to maintain current process variability.	• Adjustments will be continuously needed. • Consistent OCAPs sometimes lead to finding the actual special cause. • Process variability improved only if special cause was found. • Frequency of OOC events is reduced if the special cause eventually found and removed. • Since the OCAP adjustment is the same, you very quickly find out if it is effective. • *If special cause is not found, variability remains the same and the frequency of OOC events occurred at the same rate.*

OCAP Strategy	Comments	Observed OCAP Effects
If any sample OOS, scrap the OOS item, no other action.	• Bad	• This is product control, not process control. • Process variability tends to degrade if it is a new special cause. • Process variability remained the same if the special cause was always there. • Special Cause rarely found. • Frequency of OOC events continues at same rate if the special cause was always there. • Frequency of OOC events increased if it was a new special cause.
If any sample is OOS, scrap the OOS item and adjust the process.	• Bad	• This is product control, not process control. • Process variability tends to degrade. • Special Cause rarely found. • Frequency of OOC events tends to increase.
Different shifts/personnel allowed to implement different OCAPs.	• Ugly	• Often the result of "Call Engineering" OCAP. • Process Variability always seemed to degrade. • Often caused yield busts. • Special causes never seemed to be found. • Frequency of OOC events tends to increase.
Engineering authorizes an unproven process adjustment.	• Ugly	• Often the result of "Call Engineering" OCAP. • Process variability tends to degrade. • Often causes yield busts. • Frequency of OOC events tends to increase.

OCAP Strategy	**Comments**	**Observed OCAP Effects**
If samples are all-in-spec, no action required, reviewed OOC event(s) at weekly engineering SPC review meeting versus immediate action.	• Ugly	• Often the result of "Call Engineering" OCAP. • This is product control, not process control. • Delayed actions cause process variability degradation. • Delayed action often caused yield busts. • Frequency of OOC events tends to increase.

From the table you can see that the type of OCAP makes a difference. Some work better than others. Notice, the only OCAP that leads to process stability and variability reduction is if the OCAP involves quick response often stopping the process to investigate, locate and remove the cause. Other "Okay" OCAPs often maintain current stability, although sometimes the process stability can slowly degrade. "Bad" and "Ugly" OCAPs ensure the process stability will degrade and often causes yield busts. Also, "Bad" and "Ugly" OCAPs often invite specification control versus process control.

RECAP

How you set up the control chart is critical to maintaining process stability and capture the promise of improving the process's variability by removing special causes. It is not as simple as just plugging the data into a software program. The type of OCAPs and consistency of executing the OCAPs also makes a difference. For example, competing priorities, such as the need to meet specifications can cause properly setup charts not to work if the OCAPs focus on only meeting the specification.

In Chapter 7, I provide a comprehensive methodology for setting up and releasing a qualified well-behaved control chart to production. But first, let's explore how I discovered the *Effective Sample Size* referred to in section 5.3 in Chapter 6.

6. DISCOVERING THE EFFECTIVE SAMPLE SIZE (n_{eff})

In Chapter 5, I stated that the #1 control limit setup error is that the control chart is setup with the wrong sample size. Process Engineers are often confused with this statement as they already know the sample size from the data. I suggested that adjusting the control chart for the *Effective Sample Size* (n_{eff}) was the solution. I also promised to review the exact case on how I discovered (n_{eff}) and why it works. So, here it is!

I was working as a Consulting Engineer in a semiconductor wafer fab. Working with other colleagues, we were responsible for introducing SPC methods to the organization and deploying SPC Charts across the wafer fab. One of the first processes we attempted to deploy SPC Charts was a wet etch process. The details of the process are crucial to understanding (n_{eff}).

One production lot consisted of 50 wafers. Each wafer had 208 semiconductor devices. So, each production lot consisted of 10400 production units (208 devices x 50 wafers)! The process was a Buffered Hydrofluoric Acid (BHF) etch bath. The wet process etched away a certain amount of oxide leaving a specified "contact hole" size that was measured as the CTQ response variable. Wafers were placed in the BHF bath for a specified period, rinsed, and then dried.

Prior to the etching process, wafers from the parent production lot were placed in two process wafer boats that held 25 wafers each as depicted in Figure 10a. So, each process run consisted of 2 wafer boats with 25 wafers each in the BHF bath.

Figure 10a Wafer Boats

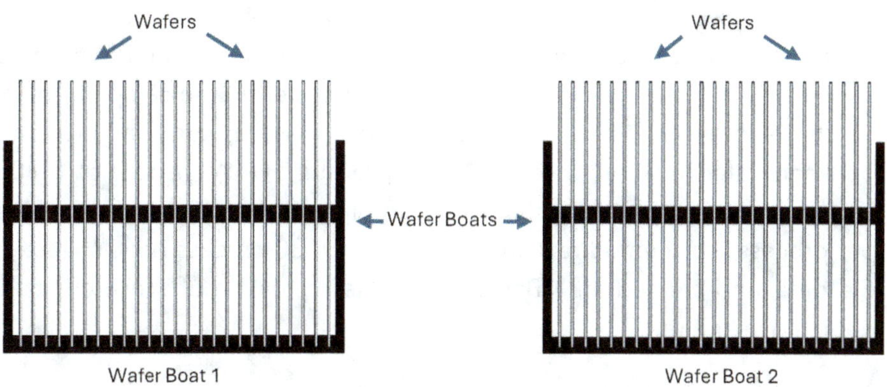

The production operator would place wafer boat 1 in the etch bath, then quickly wafer boat 2 and start a timer. Once the buzzer went off, the operator removed each wafer boat, quickly dunked each boat in a water bath which stopped the chemical reaction (the etching), then completed a rinse cycle removing all HF acid and finally put them through a dryer. Work instructions called for the operator to select 3 wafers from each boat at positions 1, 13 and 25. The operator then measured and recorded the contact hole size (CTQ) from 3 designated sites from each wafer as depicted in Figure 10b (the black squares). The sites were selected to best represent the Top, Center, and Bottom of each wafer.

Figure 10b Wafer Sampling Locations

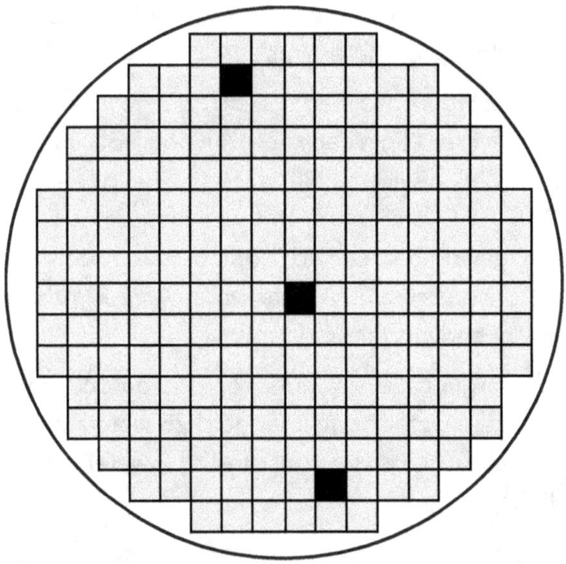

So, each process run had 6 wafers sampled, with 3 sites measured from each wafer or a total of 18 CTQ measurements for each process run. The process engineer used "rational subgroup" principles to set up the sampling plan, which we agreed with. She wanted to understand within wafer variation and between wafer variation. Given that each production lot consisted of 10400 production units (computer chips) the sampling plan of 18 computer chips per process run seemed reasonable. So, we ran 30 process runs, collected the data, and created an X-Bar, s-Chart. The recreation of these charts is depicted in Figure 11.

Figure 11 X-Bar, s-Chart (n=18)

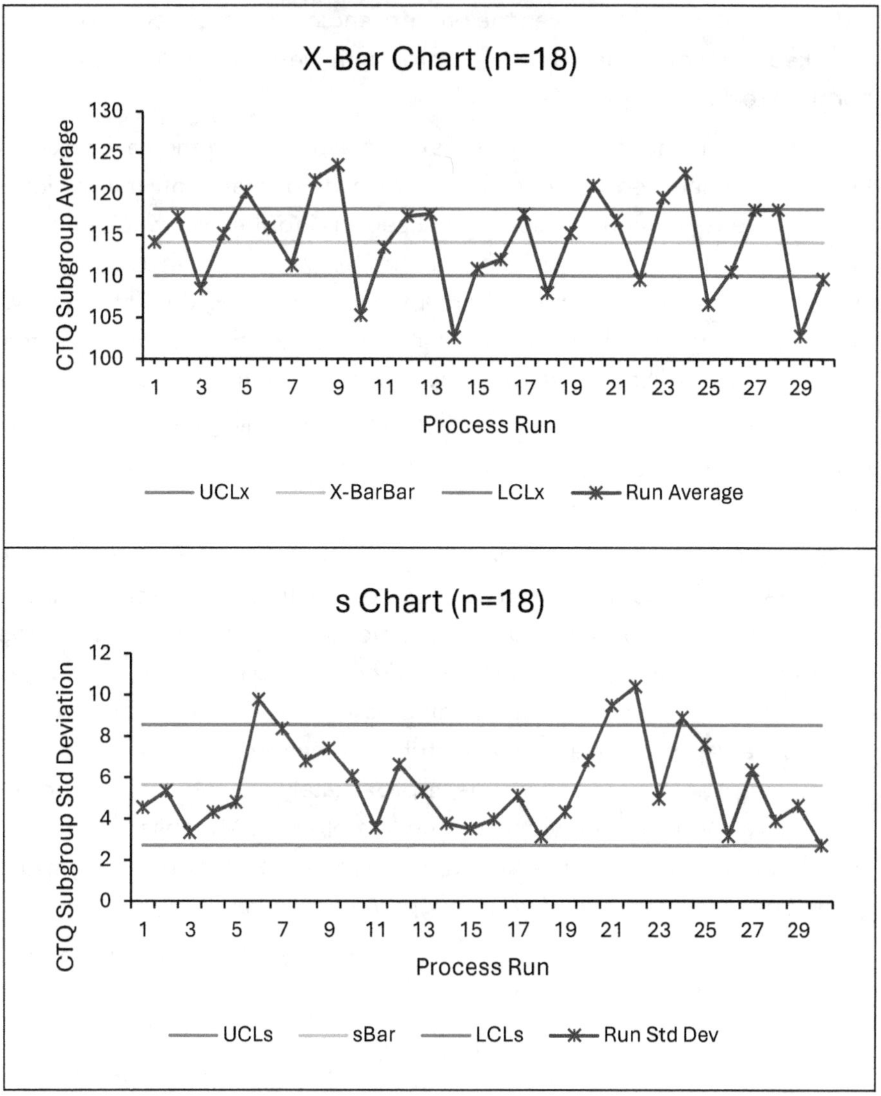

We were all surprised by the results as 18/30 runs violated special cause rule #1 (1 point out of control) - or 60% of the process runs were out-of-control! We did not even have to apply the other rules. The process engineer was quite concerned as she did not think the process was out of control. She was also concerned that she was supposed to stop the process 60% of the time to investigate the special cause. She was worried that she would not have the time to investigate given the pressure she would get from production management. She also did not know how to handle the 2nd shift, when no process engineer was on duty. Nobody else thought the process was that bad either, including me. Given the process engineer's reasonable concerns, we asked her not to deploy the control until we had conducted a more thorough review.

We did all the actions you would expect. We checked the calibration of the measurement equipment. We re-measured many of the process samples. We re-verified the constants used during the control chart setup for n=18 (A_3 = 0.718, B_3=0.482, and B_4=1.518). We re-calculated the control limits and replotted the control charts by hand and using our software. We plotted a histogram of all the raw data and concluded it was a normal distribution. All indications were that we did it according to instructions.

I then started reviewing the core principles of statistical process control methods and the *Central Limit Theorem.* Two key principles captured my attention:

1. For a sufficiently large sample size from a population, the mean (average) of all samples from the same population will be approximately equal to the mean (average) of the population. My observation: *The histogram of the data seemed to confirm this assumption.*
2. All independent and random samples drawn from a population will follow an approximate normal distribution pattern, with the variance of the sampling distribution will be approximately equal to the variance of the total population divided by the sample size. My observation: *The histogram of averages seemed to be a normal distribution, but the width of the histogram seemed much larger than expected.*

The second key principle (#2) was particularly interesting to me as it translates into the following equation:

Equation 1

$$\sigma_{\bar{X}} = \frac{\sigma_i}{\sqrt{n}}$$

where σ_i *= standard deviation of all individual measurements*

and $\sigma_{\bar{X}}$ *= standard deviation of the sample averages*

In words, it states that the standard deviation of the sample averages should equal the standard deviation of the sample individuals divided by the square root of the subgroup sample size (n).

Let's apply these principles to this example. The sample average (μ) for all 540 readings (30 runs x 18 samples per run) is equal to 115. The standard deviation (σ) for all the individual measurements = 8. Figure 12 shows 4 different *theoretical* normal distributions. The grayed distribution (Individual Measurements) represents the *theoretical* probability density distribution for all individuals with a mean of 115 and standard deviation of 8. The dotted line distribution shows the *theoretical* probability density distribution of sample averages when 2 samples (n=2) are randomly drawn from the individual population, and the probability density function is plotted. The dashed line shows the *theoretical* sample average distribution when 6 samples (n=6) are drawn from the individual population. The solid line shows the *theoretical* distribution when 18 samples (n=18), our case, are drawn from the population. Notice as the sample size increases the sample average distribution appears to get tighter. The larger the sample size, the *tighter* the distribution of averages. This is a pictorial representation of key principle #2. It is this principle that drives the Table of Constants for Control Charts (see Appendix Chapter 15). The larger the sample size, the tighter the distribution of averages, which causes the control limits to be tighter.

Figure 12 Distributions for Various Sampling Plans

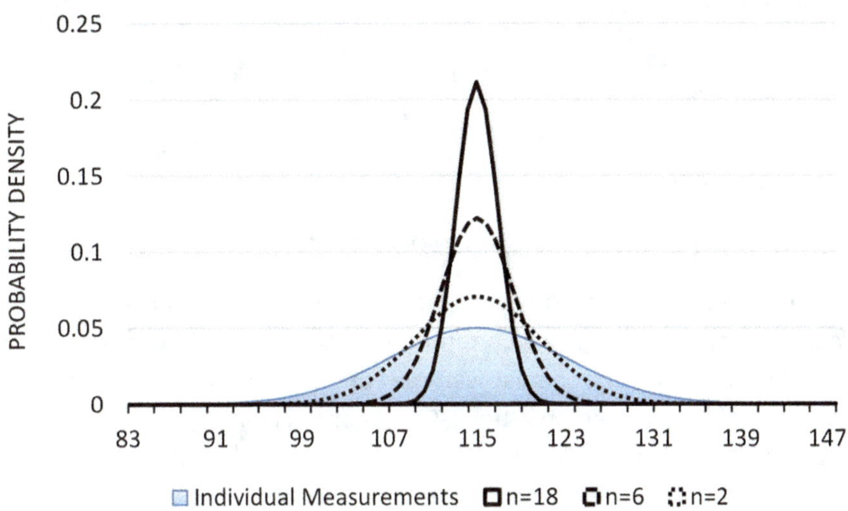

Let's confirm this principle. Table 11 lists the control limits for an X-Bar, s-Chart based on sample size. You can see our case at n=18 (6 wafers x 3 sites per wafer), has the tightest control limits. I also calculated the control limits where n=6. Recalling that in this case, we sampled 6 wafers per process run. Finally, I calculate the control limits where n=2, remembering that each process run consisted of 2 wafer boats. Notice that in all cases the control limits correspond close to the tail of each distribution in Figure 12.

Table 11 Control Limits as a Function of Sample Size

Population Average (μ)				115				
Population Standard Deviation (σ)				8				
n	A_3	B_3	B_4	C_4	LCL_x	UCL_x	LCL_s	UCL_s
18	0.718	0.482	1.518	.9854	109	121	4	12
6	1.287	0.030	1.970	.9515	105	125	0	16
2	2.659	0	3.267	.7979	94	136	0	26

With this analysis, I confirmed that our variation of sample averages was nowhere near the theoretical average variation for a sample size of 18, which has the tightest control limits! So, our process was not following key principle #2!

I distinctly remember the Aha moment! The day after I did this analysis, I arrived at work, got my cup of coffee, and approached my colleagues. I walked up to them, showed them my analysis, and suggested: "Maybe we should calculate n instead of assuming n!" They looked at me, initially confused, but after further discussion they were open to the idea. So, I rearranged the equation:

Equation 2

$$n_{eff} = \frac{\sigma_i^2}{\sigma_{\bar{X}}^2}$$

where σ_i = standard deviation of all individual measurements

and $\sigma_{\bar{X}}$ = standard deviation of the sample averages

In this case, I plugged our data into the equation and solved it for the *Effective Sample Size* (n_{eff}), which turned out to be 1.98! This of course was not even close to the actual sample of 18 measurements per run! Nevertheless, I re-created an X-Bar, s-Chart control limits for n=2, rather than n=18, which is depicted in Figure 13 Control Chart with n=2.

Figure 13 Control Chart with n=2

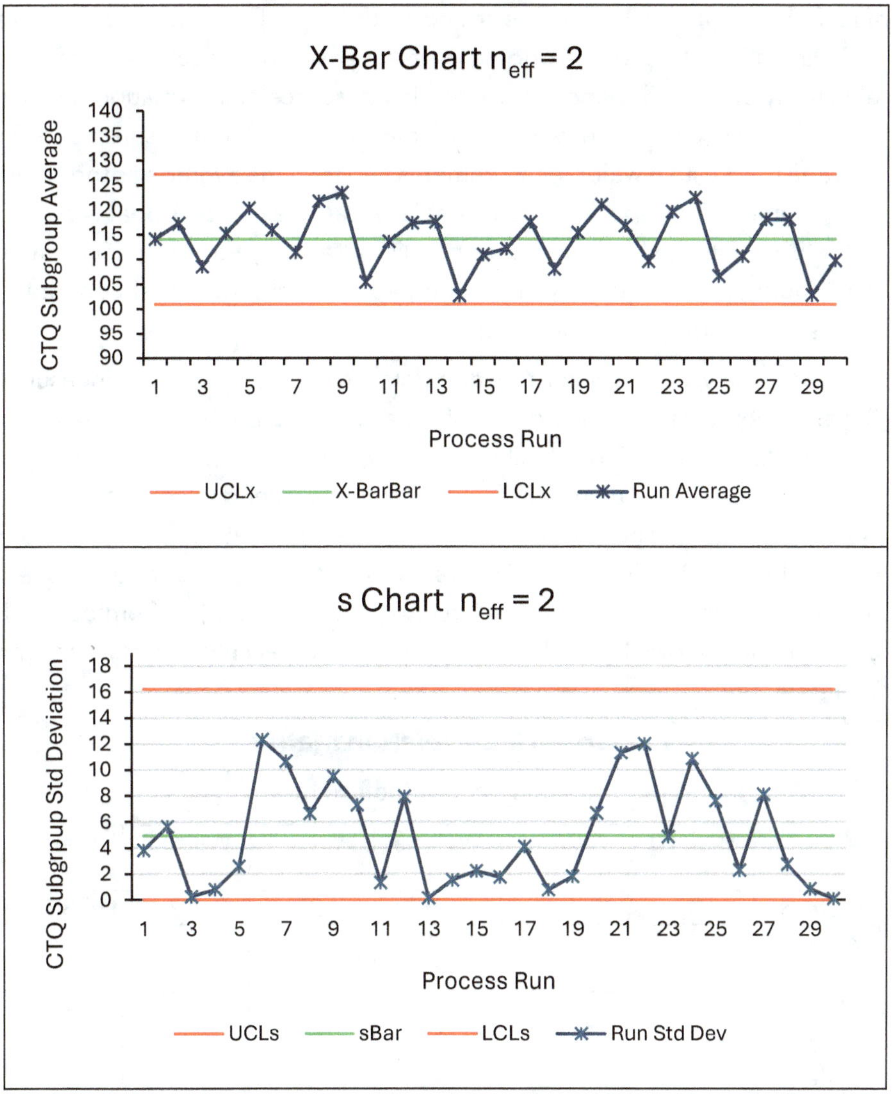

What a difference! The same data now shows no points out of control on the X-Bar or S-Chart! This control chart indicates that the process is stable!

This led me to believe that the fundamental assumption that the samples taken from the process were random and independent was not true, meaning that some of individual measurements within the subgroup samples are correlated with each other. The next challenge is to find out why (n_{eff}) = 2 and which 2 samples we should use, i.e., are the most uncorrelated.

First, I examined the possible correlation of within wafer sampling. Recall, we sampled 3 sites per wafer. I used Equation 2 on the within wafer dataset and found (n_{eff}) = .97, which is rounded up to 1. The implication of this calculation is that out of the original 3 sites measured on each wafer, there is effectively only 1 independent (uncorrelated) source of information on each wafer. So, we could use either 1 of the within wafer measurements or average the 3 within wafer measurements and treat the average as the single measurement. This appeared reasonable because as a wafer boat is placed in the BHF bath, just by the nature of a liquid equally covering each wafer, you would expect equivalent results on each wafer. Additional correlation analysis supported this conclusion.

Next, I analyzed the wafer-to-wafer dataset. Recall, production measured 3 wafers per boat (2 boats) or 6 wafer per process run. Using Equation 2, revealed (n_{eff}) = 2.03. Table 12 shows the correlation matrix for this case. This least square regression analysis for the between wafer dataset shows the R^2 correlation coefficients. From the Table it appears that Wafers 1, 2 and 3 (high value of R^2) all from boat 1 correlate with each other, but do not seem to correlate very well with Wafers 4, 5 or 6 (low value of R^2) all from boat 2. It also appears that Wafers 4, 5 and 6 correlate with each other, but not with Wafers 1, 2 or 3.

Table 12 Correlation Matrix

R^2 Correlation Coefficient

	Wafer 1	Wafer 2	Wafer 3	Wafer 4	Wafer 5	Wafer 6
Wafer 1	1	.97	.96	.25	.12	.37
Wafer 2	.97	1	.95	.27	.22	.24
Wafer 3	.96	.95	1	.31	.34	.35
Wafer 4	.25	.27	.31	1	.96	.97
Wafer 5	.12	.22	.34	.96	1	.95
Wafer 6	.37	.24	.35	.97	.95	1

From a control chart point of view, we want to select the most uncorrelated samples, this means that we only need 1 measurement from any wafer from Boat 1 and 1 measurement from any wafer from Boat 2! Or take the average of all measurements within boat 1 and treat it as 1 reading, do the same for boat 2. This confirms that the *Effective Sample Size* (n_{eff}) *=2* for purposes of the control chart even though actual measurements taken were 18 measurements per process run. How could this be?

Notice I am introducing the term Effective Sample Size (n_{eff}) in the equation. The name is derived from the concept that the original sampling plan may contain correlated measurements. Correlated samples violate a fundamental principle of the Central Limit Theorem where samples drawn from the process are ***assumed to be random and independent.*** *Using a sample size that contains correlated data causes the control limits to be tighter than they should be resulting in false OOC events. The goal during control chart setup is to select the most uncorrelated or the most "Effective Sample Size" for the control chart.*

After this analysis, and before we released the control chart to production, we spent some time observing the process to understand if there was any physical meaning for the *Effective Sample Size (n_{eff}) = 2.* We quickly noticed that there was an inconsistent delay time after the operator placed boat 1 in the BHF bath and before the operator placed boat 2 in the BHF bath, short but noticeable. We also noticed that once the timer buzzer went off, the operator would promptly remove the first boat and place it in the water quenching bath, then return and pull the second boat and place it in the water quenching bath, short delay but noticeable. Our conclusion, there was minor, yet noticeable total etch time (amount of time each boat was in the BHF bath) difference between boat 1 and boat 2. So, the difference in etch time for boat 1 compared with boat 2 caused (n_{eff}) = 2. The operator's timing delays of wafer boat transfers represents the random and independent actions that drove the biggest process variation. Eventually this finding led to installing robot arms to place the wafers boats in the etch bath as well as to remove wafers from the etch bath at the specified time. This action caused consistent placement and removal of boats in the buffered HF bath, i.e., removing that source of variation. As you would expect, the total contact hole size variability improved!

I approached the process engineer and advised her that the control chart using the *Effective Sample Size (n_{eff}) = 2* should be released to production. She was incredibly happy with the control chart as it did not show any OOC events. She felt confident that when an OOC event occurred, she could stop the process and investigate. Production was also very happy. Not only did they have a useful control chart as the process would not be stopped very often, but they only needed to take 2 CTQ readings per process run! That would save them a lot of production time. The process engineer was somewhat skeptical about reducing the sampling plan. We finally compromised and all agreed to temporarily continue taking 18 samples until our confidence grew. But we kept the control limits calculated for (n_{eff}) =2. After six months, as everyone's confidence grew, the sampling plan was finally reduced to 2 measurements per process run. Everyone was happy!

This case was a remarkable learning experience. The creditability of this technique is derived from the ability to relate (n_{eff}) calculation with physical characteristics of the process. We continued our effort and analyzed over 200 processes in this wafer fab using these techniques in this wafer fab which continued to confirm our approaches. We always seem to find a physical meaning for (n_{eff}) relating directly to how the process was performed. For example, a furnace process which had 3 heating zones (n_{eff}) was determined to be 3, while a 2-zone furnace process (n_{eff}) = 2, a single wafer process (n_{eff}) = 1, etc.

We judged these learnings to be so important, we authored a paper "Effective Sample Size - A Method for Determining Control Chart Sample Sizes for Batch Processes" and published it and delivered it at the Electro Conference in May of 1990! As indicated in Chapter 5 Case 1 - The Task Force, once I deployed this technique, I never had a problem with this organization implementing SPC Charts with OCAPs that stopped the process and investigated the special cause for an OOC event.

Even better, I continued using this technique throughout my career and found it to be applicable in many other industries. As a Quality Executive I was responsible for my organization's and our Supplier's SPC programs. As Founder and President QLRFocus LLC, I audited and reviewed many SPC programs. The majority of these SPC Programs were not associated with a semiconductor wafer fab, yet I found the *Effective Sample Size* (n_{eff}) method to be equally applicable and effective during control chart setup. The extensive applicability of the *Effective Sample Size* (n_{eff}) method with many industries is one of the reasons why I chose to author this book!

RECAP

Setting up a control chart based on CTQ samples that are correlated with each other (i.e., not random, or independent) will lead to control limits that are too tight. Control limits that are too tight will cause false OOC events and potentially unnecessarily disrupt production. False OOC events cause ineffective OCAPs and starts the *"domino effect."* Use the *Effective Sample Size* (n_{eff}) method to determine sample size that should be used to calculate control limits. Use correlation studies to determine the most uncorrelated CTQ samples from the original sampling plan. If you select the most uncorrelated sites for the control chart, you can stop measuring the unneeded correlated CTQ samples. Alternately, you may continue to measure and record the original sampling plan but use the *Effective Sample Size* (n_{eff}) number to setup your control limits.

Finally, expect skepticism when first deploying this method. Do not be afraid to keep the original sampling plan until confidence grows. But be sure to use (n_{eff}) to setup the control chart. Once confidence grows, feel free to reduce the sampling plan.

7. HOW TO SETUP AND QUALIFY A CONTROL CHART

In preceding chapters, I provided a quick SPC Overview and then discussed with some detail many reasons why SPC Control Charts do not work. Reasons range from setup errors to competing priorities etc. I also provided some insight into countermeasures that if implemented will release effective SPC charts to Production. Process Engineers who implemented these methods were rewarded with stable processes. If they ran OCAPs that exposed special causes, and subsequently removed these special causes, their overall process variability improved. For those organizations that consistently implemented these methods across their factory, their "SPC for Show" Programs changed to "SPC for Improvement" Programs that achieved operational excellence.

In this chapter, I am pulling all these learnings together. These instructions pertain to the initial setup and qualification of a variable control chart for a process that is ready to be released to production. You should find that these methods are consistent with quality system expectations of ISO 9001, AS 9100, TS 16949 as well as the APQP framework.

The 16 steps on how to setup and qualify a control chart before it is released to production are:

1. Identify the Process, Process Equipment, and the CTQ.
2. Verify the Readiness of Measurement Equipment.
3. Verify Work Instructions and Production Operators are Certified.
4. Define the Initial Sampling Plan (n_{init}).
5. Setup Trial Control Limits for Pilot Runs.
6. Conduct 30 Process Runs and Measure the CTQ.
7. Analyze the Dataset for the Effective Sample Size (n_{eff}).
8. Determine the Most Random and Independent Samples.
9. Decide if the Ongoing Sampling plan Should be Reduced to (n_{eff}).
10. Construct a Control Chart using (n_{eff}) to set Control Limits.
11. Analyze and Remove Obvious Lurking Variables.
12. Calculate the *Stability Index* (S_i) and Apply the 85% Rule.
13. Define Quick Response OCAPs.
14. Define OSAPs for Product CTQs (when necessary).
15. Enable Out-of-Control (OOC) Rules with Quick Notifications.
16. Release the Qualified Control Chart to Production.

Once you successfully complete these steps, you can be confident that you have released a *well-behaved* control chart for a process that is *reasonably stable* to production absent of lurking variables with properly sized control limits. Process monitoring and improvement begins!

Let's review each of these 16 steps in more detail.

7.1. Identify the Process, Process Equipment, Process Runs, and the CTQ

Seems simple enough to identify the process, the process CTQ, the product CTQ (if applicable) and the process equipment. However, there are key issues to be aware of during this step.

First, be aware of process CTQs and product CTQs differences. Recall from Chapter 4 that some CTQs can be both product and process CTQs. If you use a CTQ that is both a product and a process CTQ, it invites competing priorities which can start the "*domino effect*" leading to ineffective control charts. It is not that you cannot use CTQs that are both a product and process CTQ. However, separate them whenever possible. As suggested in Chapter 4, if you separate them, it makes it easier to construct OCAPs for Process CTQs and OSAPs for Product CTQs. Finally, make sure you do not invent process specifications for Process CTQs when they are not needed.

Second, understand the relationship between production lots, production units and process runs (Chapter 5.1). Be sure when you prepare the data for analysis and control charting (especially when you are using SPC production software), you are plotting subgroups by process run data and *not* production lot and/or production unit data. Unless, of course, if they are the same.

Third, understand equivalent process equipment. If you have multiple pieces of equipment running the same process, for the purpose of setting up the first control chart for the selected process, identify and select the primary piece of equipment. This will eliminate the #1 cause for *Lurking Variables* (Chapter 5.4). This is particularly important if the processes setup on equivalent equipment are not "Copy Exact" but are "Copy Similar."

Fourth, Product CTQs are often measured after several process steps (see Lurking Variable Case Study in 7.11), which complicates control chart setup. The methods in this chapter can address these situations, but it is better to choose a Process CTQ for a single step.

In the next chapter (8), I provide additional guidance on how to add equivalent equipment ("Copy Exact" and "Copy Similar") to a control chart or whether they need to remain separated.

7.2. Verify the Readiness of Measurement Equipment

Ensure your measurement equipment are ready as suggested in Chapter 5. Perform Calibrations, Gage R&R, or MSA studies before starting SPC setup runs, depending on your situation.

Calibration ensures that a measurement device is providing accurate readings by comparing it to a known standard. It involves adjusting the measurement device to eliminate any deviation from the standard. It focusses on accuracy and precision of the measurement device itself. The outcome is a calibrated device that can produce accurate measurements. It is best if only one measurement device and only one operator will be used during control chart setup runs. A suitable standard to reference is ISO/IEC 17025.

Gage R&R studies assess the measurement system's ability to consistently produce the same results under the same conditions by multiple operators. It involves repeated measurements on the same measurement device by different operators to evaluate repeatability (consistency of the same operator) and reproducibility (consistency among different operators). The focus is the measurement system's overall variability, including the device, operators, and methods. The outcome is the identification of sources of variation and determination of whether the measurement system is suitable for its intended use. It is best if one measurement device is used by multiple operators during control chart setup runs.

Measurement Systems Analysis (MSA) is a more comprehensive approach evaluating the quality and reliability of the entire measurement system. The components include calibration, Gage R&R, linearity, stability, bias, and other studies to assess the various measurement equipment and operators that will be used for SPC setup runs. The focus is the broadest as it includes all measurement devices, operators, methods, and environmental factors. The outcome ensures the entire measurement system that will be used during control chart setup produces reliable and consistent data by identifying and addressing all sources of variation. It is best used if multiple measurement devices are used by multiple operators during control chart setup runs. Suitable standard is AIAG MSA Manual.

7.3. Verify Work Instructions and Production Operators are Certified

This is a core principle of most quality systems (such as those defined in ISO 9001, TS 16949, AS 9100, ISO 13485 quality system standards etc..) that clear work instructions are written, approved, and released to production personnel. Verifying work instructions should include the process work instructions and the work instructions for any measuring equipment. In addition, all production personnel should be fully trained and certified before they run the process or take measurements needed for the control chart setup. The adequacy of the work instructions and training certification can usually be assessed by an audit performed by your Quality Department. Any findings should be addressed before control chart setup runs begins.

Production personnel are usually Operators but can sometimes include Technicians and rarely include Process Engineers. Examine the work instructions to determine who should be running the process. *You want to make sure that only authorized (trained and certified)* ***production personnel*** *conduct the control chart setup process runs as specified in the work instructions.* For example, if work instructions indicate operators will normally run the process, do not accept process runs completed by engineering. The worst thing you can do is have Process Engineering conduct all the control chart setup runs, analyze the data, and then release the chart to production. From experience, you can guarantee many false OOC events once production runs the process.

7.4. Define the Initial Sampling Plan (n_{init}).

The initial sampling plan for each process run subgroup is intended to be only used during the initial 30 process runs used to setup the control chart. You should take into consideration key equipment features and other anticipated variation within and between production units, within and between production lots; and, within and between process runs. Given that you are going to analyze this data for the *Effective Sample Size* (n_{eff}), I encourage taking *more samples* that you will likely need after the *Effective Sample Size* (n_{eff}) analysis. This is particularly true if this is the first time you are analyzing the process on the primary process equipment. As discussed in Chapter 6, the *Effective Sample Size* (n_{eff}) analysis will reduce the final sampling plan to those samples that are the most random and independent from each other. You can then choose to reduce the original sampling plan (n_{init}) to (n_{eff}) for ongoing production. However, no matter what sampling plan you use in production, you will be using (n_{eff}) to setup the control limits.

For example, one equipment feature typical of semiconductor thermal ovens, such as large multi wafer lot gate oxide growth ovens, are multiple heating zones. My experience suggests that the *Effective Sample Size* (n_{eff}) analysis often revealed that the final sampling plan should include 1 sample from each heating zone.

Note, if this is a repeat control chart setup for "Copy Exact" equivalent equipment, my experience suggests that you can immediately use the (n_{eff}) previously determined.

However, if this is a repeat control chart setup for "Copy Similar" equivalent equipment, you **should not** use (n_{eff}) previously determined for the process from other equipment. You should treat this rational subgrouping definition stage as if it was a completely new process. My experience suggests that *sometimes* (n_{eff}) is the same for "Copy Similar" equivalent equipment, but often it is different. It is best to determine this during the setup runs.

7.5. Setup Trial Control Limits for Pilot Runs

I often got asked "How do I control my Pilot Runs when I do not have any data? Especially if you want me to collect 30 process runs before I calculate control limits." So, I developed this approach to use *Trial Control Limits* during the initial Pilot runs. Data from these Pilot Process runs will ultimately be analyzed for (n_{eff}) and determine final control limits as described in Section 7.10.

It is also important to define Pilot Run OCAPs in case a process run violates a special cause rule. These OCAPs should investigate reasons for detected shifts in the process. If adjustment OCAPs are used to permanently move the process distribution during Pilot Process runs, re-start the Pilot Runs for the purpose of calculating control limits and qualifying control charts.

This approach involves using C_{pk} formulas described in Chapter 10 and formulas from the Central Limit Theorem covered in Chapter 6, along with some key assumptions.

First, consider the C_{pk} equation:

$$C_{pk} = \left(\frac{USL-\mu}{3\sigma_i}\right)$$

Where: USL = Upper Specification Limit
LSL = Lower Specification Limit
u = Average of all Individual Measurements
σ_i = Standard Deviation of all Individual Measurements

Second, rearrange the equation to solve for σ_i.

$$\sigma_i = \left(\frac{USL-\mu}{C_{pk}}\right)$$

Next, we take advantage of the Central Limit Theorem formulas used to calculate the standard deviation of averages:

$$\sigma_{\bar{x}} = \frac{\sigma_i}{\sqrt{n}}$$

This approach can be applied whether production specifications for Product CTQs have been established or not, as well as for Process CTQs lacking specifications. In situations where specifications are not set, substitute the Maximum Acceptable Value (MAV) for the USL and the Minimum Acceptable Value (MIN) for the USL. The MAV and MIN values for Product CTQs are often determined with assistance from Marketing. For Process CTQs, MAV and MIN values are typically set by Process Engineering.

Finally, we need to make a key C_{pk} assumption. I recommend a C_{pk} assumption between 1.0 and 1.33. I always found that assuming $C_{pk} > 1.33$ during Pilot Runs was too aggressive as it will calculate too tight Trial Control Limits.

Let's consider an example where the USL = 2400 and the LSL = 1600. Also, the initial sampling plan (n_{init}) as determined by methods described in 7.4 is equal to 5. From these assumptions we can calculate the expected process average is equal to (USL + LSL)/2. Since the sample size = 5, using the recommendations described in *Table 3* Recommended Chart Types to Use, we implemented an X-Bar, s Chart. Look up the A_3, B_3 and B_4 constants from Table 36 Constants for X-Bar, s Charts and proceed with the calculations.

Table 13 Trial Limits for Product CTQs with USL & LSL

USL (or MAV):	2400	Sample Size (n_{init}): 5
LSL (or MIN):	1600	
	Formula	Example Calculation
Calculated Average μ:	$\mu = (USL + LSL)/2$	$2000 = (2400 + 1600)/2$
Calculated σ_i:	$\sigma_i = (USL - \mu)/C_{pk}$	$301 = (2400\text{-}2000)/1.33$
Calculated $\sigma_{\bar{x}}$:	$\sigma_{\bar{x}} = \dfrac{\sigma_i}{\sqrt{n_{init}}}$	$135 = (301/\sqrt{5})$
<u>X - Bar Chart</u>		
Trial $UCL_{\bar{x}}$:	$UCL_{\bar{x}} = \mu + (A_3 \times \sigma_{\bar{x}})$	$2192 = 2000 + (1.427 \times 135)$
Center Line:	μ	2000
Trial $LCL_{\bar{x}}$:	$LCL_{\bar{x}} = \mu - (A_3 \times \sigma_{\bar{x}})$	$1808 = 2000 - (1.427 \times 135)$
<u>S - Chart</u>		
Trial $UCL_{\sigma_{\bar{x}}}$:	$UCL_{\sigma_{\bar{x}}} = B_4 \times \sigma_{\bar{x}}$	$281 = (2.089 \times 135)$
Center Line:	$\sigma_{\bar{x}}$	135
Trail $LCL_{\sigma_{\bar{x}}}$	$LCL_{\sigma_{\bar{x}}} = B_3 \times \sigma_{\bar{x}}$	0
<u>Constants</u>	$A_3 = 1.427$	$B_3 = 2.089$ $B_4 = 0$

7.6. Conduct 30 Process Runs and Measure the CTQ

Now you are ready to run the process and collect the data per the originally defined sampling plan (n_{init}) for each process run subgroup. I recommend a minimum of 30 process runs (not production lots) during the control chart setup. Remember, per section 7.3, only certified production personnel should conduct process runs and take sample measurements. The goal for these 30 process runs is to find and determine the expected long-term variability across all production personnel and shifts. This translates into more appropriate control limits and often immediately identifies the most prevalent special causes.

If you are a multiple shift operation, you have a couple of choices. My recommendation is that you conduct the 30 process runs using all the production personnel for your primary shift, ensuring that all certified production personnel for that shift complete some of the process runs. Once the control chart has been setup and released to production, conduct 10 process runs by the next shift and plot on them by shift sequentially on the control chart. Repeat until all shifts are done. If no control chart rules are violated, you may begin plotting data from all the passing shifts on the same control chart. If a rule is violated - you found a special cause! Investigate, identify the reason(s), and remove the special cause!

Alternatively, if the first choice is too inconvenient, conduct the 30 or more process runs equally across all shifts using all certified production personnel. Remember, you can always run more than the recommended 30 runs if needed and that you will analyze the data for the "Lurking Variables" before you release it to production. Again, if a rule is violated - you found a special cause! Investigate, identify the reason(s), and remove the special cause!

The reason I am emphasizing this is because you want to capture variability across all shifts, from all certified production personnel. My experience suggests that the vast majority of OOC events after the initial release of control charts are caused by inconsistent execution of the process by production personnel, inadequate work instructions, work instructions that allow on-demand untested process adjustments, or inconsistent adjustments made between shifts and/or different personnel.

For example, even though operators are certified, some operators do not follow instructions, they run the process from their personal knowledge. Also, be on the watch for work instructions that instruct operators to "Call Engineering" when uncertainty happens, or when production unit(s) fall out-of-spec. Process Engineers arrive and may authorize adjustments. Often the authorized adjustments are inconsistent, untested, and can vary by Process Engineers. These "adjustments" almost guarantee the release of unstable processes. It is helpful if someone, like a Quality Engineer, observes, verifies, and audits control chart setup runs. They can verify that production personnel ran the process according to work instructions, or whether the work instructions are inadequate, or if too many adjustments are needed to run the process.

All of these are examples of special causes that can appear during control chart setup. If you find them early during setup, this is the best and easiest time to find them - and take action to remove them!

7.7. Analyze the Dataset for the Effective Sample Size (n_{eff})

Now you should have a (n_{init}) dataset for at least 30 process runs. The next step is to analyze the dataset for the *Effective Sample Size* (n_{eff}). Recall in section 7.4, I encouraged you to define the subgroup to capture as much variation as possible during the control chart setup runs. This means that during the setup runs, you could and should be over sampling. Why is oversampling a good strategy during control chart setup runs? It gives you the best chance of accurately determining *Effective Sample Size* (n_{eff}); and you can reduce the ongoing sampling to *Effective Sample Size* (n_{eff}) when you release the control chart to production.

Recall from Chapter 6, that the Central Limit Theorem (which control charts are based on) assumes that samples drawn for purposes of setting up a control chart *independent* and *random*. If this is true then samples drawn from the population will follow an approximate normal distribution pattern, with all sample variances being approximately equal to the variance of the population (individual measurements) divided by the sample size. This is true even if the population of all individuals is not a normal distribution. This relationship is depicted in Equation 3.

Equation 3

$$\sigma_{\bar{X}} = \frac{\sigma_i}{\sqrt{n}}$$

where σ_i *= standard deviation of all individual measurements*

and $\sigma_{\bar{X}}$ *= standard deviation of the sample averages*

This means, if you assume the samples are *random* and *independent*, and they are not, you will calculate incorrect control limits (see Chapter 6). Most of the time the control limits will be calculated too tight and cause false OOC events.

Recall from Chapter 6, that I rearranged the equation to solve for n. Further, I renamed n to be the *Effective Sample Size* (n_{eff}) as depicted in Equation 4.

Equation 4

$$n_{eff} = \frac{\sigma_i^2}{\sigma_{\bar{X}}^2}$$

where σ_i = standard deviation of all individual measurements

and $\sigma_{\bar{X}}$ = standard deviation of the sample averages

So, now use Equation 4 to calculate the *Effective Sample Size* (n_{eff}).

Consider a case where we have 30 process runs with 6 samples measured for each process run for a total of 180 sample measurements. First calculate the standard deviation for all samples taken, this will be (σ_i). Next calculate the average for each process run, this will be ($\bar{X}_i$). You will have 30 process run averages. Finally, calculate the standard deviation for the 30 sample averages, this will be ($\sigma_{\bar{X}}$). Plug it in the formula above and calculate (n_{eff}).

To demonstrate this, I included 2 key examples in Appendixes located in Chapter 15.

Appendix 15.3 is an example dataset with samples that are random and independent. Notice calculations at the bottom of the table. The *Effective Sample Size* (n_{eff}) is calculated to be 6.04, which is the very closed to the actual sample size of 6 samples measured for each process run. In this case, you would proceed with the control chart setup using all 6 samples per run, i.e., proceed to step 7.9.

Appendix 15.4 is an example dataset with samples correlated samples. Notice the calculations at the bottom of the table. The Effective Sample Size (n_{eff}) is calculated to be 2.21, which is not even close to the actual sample size of 6 measured for each process run. In this case, you should assume that the dataset contains correlated samples. You should complete Section 7.7 before calculating control limits.

NOTE: Since this analysis is not a normal part of SPC software, it is most often analyzed using suitable software like Excel and their Analysis ToolPak Add-in.

7.8. Determine the Most Random and Independent Samples

Now you have a calculated (n_{eff}). My experience shows that (n_{eff}) rarely equals the original sample size (n_{init}) used during the initial control chart setup runs. So, it is time to determine which samples from the original sampling plan (n_{init}) are the most random and independent. This step is particularly important if you want to reduce the ongoing production sampling plan to (n_{eff}). You want your final sampling plan to best represent the natural variation of the process. The best and fastest method I found was to create a Correlation Matrix Table listing the R^2 Correlation Coefficient by sample site, see Table 14. This analysis method is also not included in production SPC Software, so it is usually done offline using other software supporting statistical calculations like Excel.

Table 14 Correlation Matrix

R^2 Correlation Coefficient

	Sample 1	Sample 2	Sample 3	Sample 4	Sample n..
Sample 1	1	X1	X2	X3	X4
Sample 2	X1	1	X5	X6	X7
Sample 3	X2	X5	1	X8	X9
Sample 4	X3	X6	X8	1	X10
Sample n..	X4	X7	X9	X10	1

Notice there will be a diagonal of "1" i.e., Sample 1 should correlate with Sample 1, Sample 2 should correlate with Sample 2, etc. You should also see a mirror effect: Sample 1 vs Sample 2 R^2 (X1) should be the same as and Sample 2 vs Sample 1 R^2(X1), etc. If you do not see the diagonal or the mirror effect in the Correlation Matrix Table, re-check your dataset and analysis technique. The next step is to pick, using R^2, a sampling plan of the most uncorrelated samples (the lowest values of R^2) up to (n_{eff}) sample size. These are the sample measurements that will produce the most effective control chart. (See Chapter 6 for a more detailed example).

7.9. Decide if the Ongoing Sampling Plan Should be Reduced to (n_{eff})

At this point you have identified sample size (n_{eff}) needed to calculate control limits and have identified the most uncorrelated sample sites. You are now faced with a choice.

If (n_{eff}) = (n_{init}), then keep the original sampling plan and move on to the next step.

If (n_{eff}) < (n_{init}), then you have a couple of options. Option 1: You can immediately reduce the production sampling plan to (n_{eff}) and only measure, collect and plot the data corresponding to the most uncorrelated sites identified by your correlation matrix analysis. Production is usually very happy with this option! Option 2: You can temporarily collect process run data per the (n_{init}) sampling plan until you have sufficient confidence, then once you are confident, reduce the sampling plan. This is usually a good option when first deploying this effective sample size technique within an organization as it helps generate confidence with this technique. Option 3: you can just continue to use the (n_{init}) sampling plan. No matter which option you use, be sure to use (n_{eff}) for calculating the control limits.

However, if you do not create and analyze for (n_{eff}) using a correlation matrix to determine the most uncorrelated samples, you must use Option 3, i.e. continue to use the (n_{init}) sampling plan.

You should now analyze the process to determine if there is any physical meaning for the *Effective Sample Size* (n_{eff}). Recall from Chapter 6, the creditability of this technique is derived from the ability to relate (n_{eff}) calculation to physical characteristics of many processes across multiple industries.

7.10. Construct a Control Chart using (n_{eff}) to set Control Limits

Once you have (n_{eff}), you can now construct a control chart. As previously mentioned in Chapter 3, there are a variety of control charts that you can use. Here is the good news, you only need to consider 3 types of control charts - even if the histogram of individual samples does not appear to be normally distributed! Recall, I recommend these chart types: *X_i & MR, $\bar{X}$ & R, and $\bar{X}$ & s* chart. Other chart types work fine if they are setup correctly. But they are more difficult to setup, and it is even more difficult to train an organization on how to use them properly. Table 15 provides guidance on which control chart you should use, which depends on n_{eff}. I am assuming that you will be using production software and not manually plotting the control chart.

Table 15 Recommended Control Chart

n_{eff}	Recommended Chart Type
1	X_i & MR
2	$\bar{\boldsymbol{X}}$ & R
>2	$\bar{\boldsymbol{X}}$ & s

If (n_{eff}) = 1, then the best chart to use is the X_i & MR chart. If (n_{eff}) = 2, you should use the $\bar{\boldsymbol{X}}$ & R Chart. Once(n_{eff}) >2, then use the $\bar{\boldsymbol{X}}$ & s chart.

For completeness, I include the formulas pertaining to each control in the following Tables 16-18.

Table 16 Key Formulas for Individuals, Moving Range Chart

$X_i, R_i = X_{Max} - X_{Min}$	$\bar{\bar{X}} = \frac{\Sigma\,(X_1 \ldots X_k)}{}$ $\overline{MR} = \frac{\Sigma\lvert X_1 - X_2\rvert + \lvert X_2 - X_3\rvert \ldots . \lvert X_{k-1} - X_k\rvert}{k-1}$ where k = number of moving range subgroups
$\bar{\bar{X}} = Centerline$ $LCL_{\bar{X}} = \bar{\bar{X}} - E_2 * \overline{MR}$ $UCL_{\bar{X}} = \bar{\bar{X}} + E_2 * \overline{MR}$ $UCL_{MR} = D_4 * \overline{MR}$ $LCL_{MR} = D_3 * \overline{MR}$	$\mu = \bar{\bar{X}}$ $\sigma = \frac{\overline{MR}}{d_2}$
Where E_2, D_3, D_4 and d_2 are taken from a Table of Constants in Appendix Chapter 15	

Table 17 Key Formulas for X-Bar, R Charts

$\overline{X_i} = \frac{\Sigma (X_1 .. X_n)}{n}$, $R_i = X_{Max} - X_{Min}$ where n = number of observations per subgroup	$\bar{\bar{X}} = \frac{\Sigma (\bar{X}_1 \ldots \bar{X}_k)}{k}$, $\bar{R} = \frac{\Sigma (R_1 \ldots . R_k)}{k}$ where k = number of subgroups
$\bar{\bar{X}} = Centerline$ $LCL_{\bar{X}} = \bar{\bar{X}} - A_2 * \bar{R}$ $LCL_{\bar{X}} = \bar{\bar{X}} - A_2 * \bar{R}$ $UCL_R = D_4 * \bar{R}$ $LCL_R = D_3 * \bar{R}$	$\mu = \bar{\bar{X}}$ $\sigma = \frac{\bar{R}}{d_2}$
Where A_2, D_3, D_4 and d_2 are taken from a Table of Constants in Appendix Chapter 15	

Table 18 Key Formulas for X-Bar, s-Charts

$\overline{X_i} = \frac{\Sigma (X_1 .. X_n)}{n}$, $s_i = \sqrt{\frac{\Sigma (X_1 - \bar{X})^2}{n-1}}$ where n = number of observations per subgroup	$\bar{\bar{X}} = \frac{\Sigma (\bar{X}_1 \ldots \bar{X}_k)}{k}$, $\bar{s} = \frac{\Sigma (s_1 \ldots . s_k)}{k}$ where k = number of subgroups
$\bar{\bar{X}} = Centerline$ $LCL_{\bar{X}} = \bar{\bar{X}} - A_3 * \bar{s}$ $UCL_{\bar{X}} = \bar{\bar{X}} + A_3 * \bar{s}$ $UCL_s = B_4 * \bar{s}$ $LCL_s = B_3 * \bar{s}$	$\mu = \bar{\bar{X}}$ $\sigma = \frac{\bar{s}}{c_4}$
Where A_3, B_3, B_4 and c_4 are taken from a Table of Constants in Appendix Chapter 15	

7.11. Analyze and Remove Obvious *Lurking Variables*

The next step is to analyze the control chart for obvious *Lurking Variables*. As previously discussed in Chapter 5, a *Lurking Variable* observed *during control chart setup* is defined as an assignable cause, failing Rules 2, 3 or 4 when applied to CTQ subgroup control chart setup data. Basically, if you fail one of these during control chart setup, there is an obvious special cause, and the best time to address it is during control chart setup. Below are 2 control charts to demonstrate.

Figure 14 No Evidence of a Lurking Variable

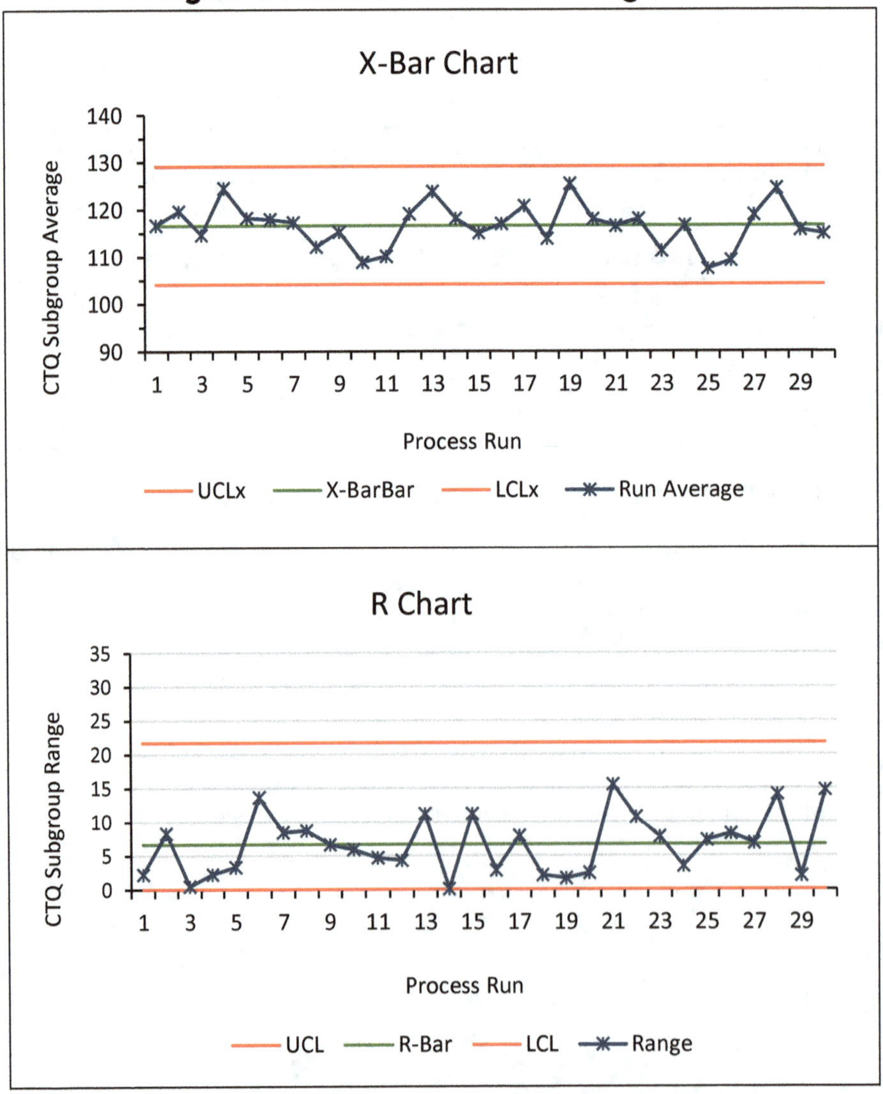

You should apply Rules 2, 3 and 4 from Figure 15 Control Chart Rules to the control chart in Figure 14. The control chart in Figure 14 has no violations of Rules 2, 3 or 4, so you should move on to the next step (Section 7.12).

Figure 15 Control Chart Rules

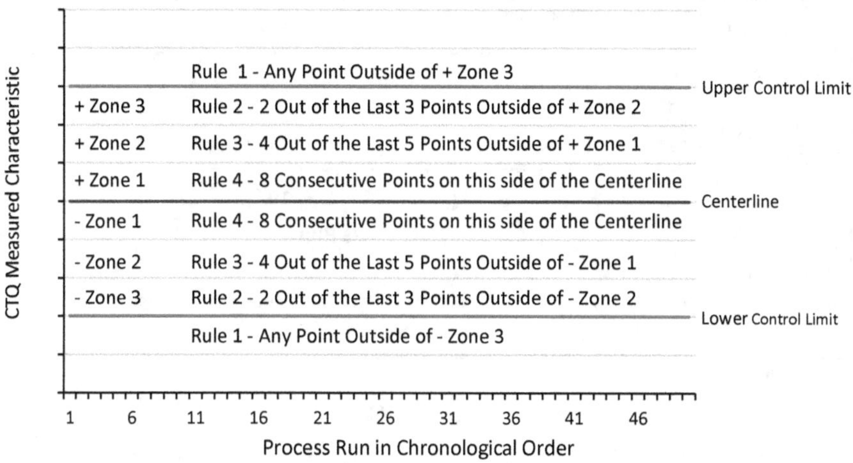

Now consider the control chart in Figure 16.

Figure 16 Lurking Variable Example

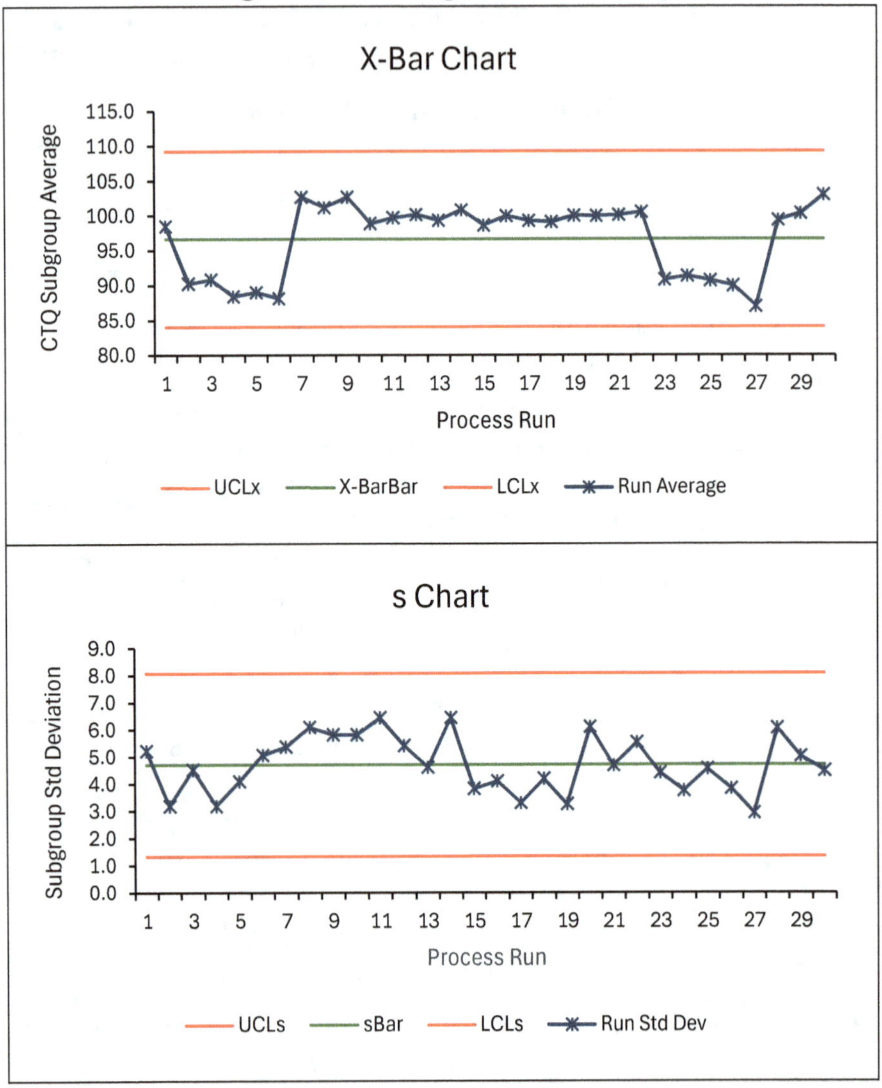

Process runs #2-6 fail Rule 3, runs #7-22 fail Rule 4; and runs #23-27 fail Rule 2. So, we have identified at least 1 *Lurking Variable* during control chart setup - maybe there are 3! The possible *Lurking Variable(s)* should be investigated and addressed before this control chart is released to production. Most of the time, the lurking variable appears on the X-Bar chart rather than the s-Chart (or R-Chart).

You should launch an investigation to determine the special cause associated with the *Lurking Variable* (these OOC event(s)). NOTE: The methods described in this section can be used to investigate any OOC event.

The special cause for the *Lurking Variables* found during setup can often be found by simply auditing the control chart setup process runs. From my experience, the most common causes for *Lurking Variables* detected *during control chart setup* were described in Chapter 5, they include: measurement error, inconsistent process execution, inconsistent or unauthorized process adjustments, subgroups plotted by production lot not by process run, and/or equivalent equipment incorrectly being plotted on the same chart. Table 18 lists examples of *Lurking Variables* found during control chart setup and the most effective corrective actions.

Table 18 Corrective Actions for Lurking Variables Detected During Setup

Lurking Variable Cause	Corrective Action(s)
Lurking Variable process runs were measured by an instrument found to give incorrect readings.	• Remove the measuring instrument from production use until MSA study is completed and the instrument is found to be equivalent to other devices. • Conduct replacement process runs, plot the new data, re-analyze for *lurking variables*. • Release the instrument back to production once it passes MSA studies indicating equivalency.
Lurking Variable process runs were all completed by an individual operator.	• Re-train and re-certify operator. • Audit the operator while they run replacement process runs. • Conduct replacement process runs, plot the new data, re-analyze for *lurking variables*.
Lurking Variable process runs are associated with specific production shift.	• Re-train and re-certify operator(s) on that shift. • Audit the operator(s) while they run replacement process runs. • Conduct replacement process runs, plot the new data, re-analyze for *lurking variables*.
Lurking Variable process runs audit revealed unauthorized adjustment(s) made before, during or after a process run.	• Adjust work instructions to ensure unauthorized adjustments are not allowed. • Re-train and re-certify operator(s). • Audit the operator(s) while they run replacement process runs. • Conduct replacement process runs, plot the new data, re-analyze for *lurking variables*.

Lurking Variable Cause	Corrective Action(s)
Lurking Variable process runs audit revealed *inconsistent* authorized adjustment(s) made before, during or after a process run.	• Adjust work instructions to ensure *identical* authorized adjustments across all personnel and shifts. • Re-train and re-certify operator(s). • Audit the operator(s) while they run replacement process runs. • Conduct replacement process runs, plot the new data, re-analyze for *lurking variables.* • (Note: with this corrective action you will find out very quickly whether the authorized adjustment is effective.)
Subgroup data was plotted as production lots instead of process runs (this is often the default plotting method of production software).	• Replot the data as *process runs* chronological order. • Conduct additional runs to meet the recommended minimum of 30 runs, if needed. • Re-analyze for lurking variables.
Lurking Variable process runs were associated with "Copy Similar" equivalent equipment. **← This is the most frequent *Lurking Variable* special cause.**	• Remove all data points associated with the "Copy Similar" equipment. • Conduct replacement process runs without the "Copy Similar" equipment, plot the new data, re-analyze for *lurking variables.* • Proceed to setup a unique control chart for the "Copy Similar" equipment.

However, there are some situations that require more rigorous analysis to determine the special cause. This is especially true if the control chart CTQ is measuring the results from multiple process steps, rather than a single process step. Consider the following case study.

LURKING VARIABLE CASE STUDY

This case study explores a Die Attach process in the semiconductor industry where a semiconductor chip is attached to a substrate or package. The process consists of Die Attach Picker selecting a chip, aligning it with the substrate or package, placing it on substrate which contains previously deposited epoxy adhesive with a controlled force. The placement is followed with a thermal curing step. A sample is selected from each process run where a Die Shear is measured as the Product CTQ. A process run often consists of placement of over 1000 die. The die shear test involves a precise shear force being applied against the edge of the die where the strength of adhesion of the die to the substrate is measured once the die breaks off and is called Die Shear, which is the response variable. The die attach process ensures both mechanical stability and heat transfer properties which directly affect product reliability - so it is a very important variable to control. The individuals control chart (Figure 17*)* shows multiple OOC events violating Rule 3.

Figure 17 Individuals and Moving Range Chart

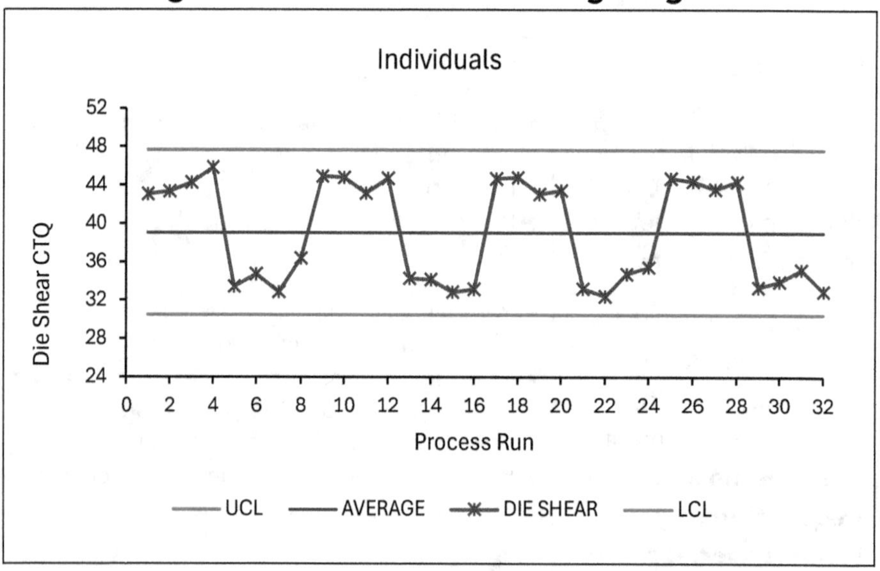

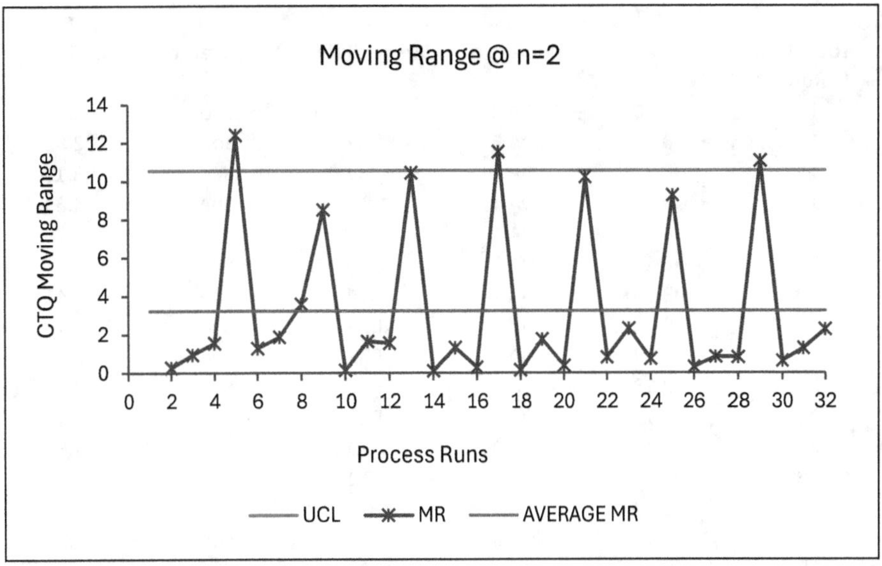

An audit of the process revealed results in Table 19 which documents 4 key variables for the 32 production runs identified in Column 1. "Die Bake" (Column 2) refers to 2 different hot plates (DB 20 and DB 18) used for the curing process step. "Die Attach" (Column 3) refers to 2 different die attach picker tools (DA 1 and DA B) which are used to place the die on the substrate. "Epoxy Batch" (Column 4) indicates that the 32 runs used 2 different epoxy adhesive batches (EB 1 and EB 2). Finally, the audit revealed that 2 different operators (Sue and Joe) performed the 32 process runs (Column 5). The response variable "Die Shear" is listed in Column 6.

Table 19 Die Attach Audit Results

1 Production Run	2 Die Bake	3 Die Attach	4 Epoxy Batch	5 Operator	6 Die Shear
1	DB 20	DA 1	EB 1	Joe	43.04
2	DB 20	DA 1	EB 1	Joe	43.31
3	DB 20	DA 1	EB 1	Joe	44.25
4	DB 20	DA 1	EB 1	Joe	45.79
5	DB 20	DA 1	EB2	Joe	33.39
6	DB 20	DA 1	EB2	Joe	34.67
7	DB 20	DA 1	EB2	Joe	32.82
8	DB 20	DA 1	EB2	Joe	36.39
9	DB 20	DA B	EB 1	Joe	44.89
10	DB 20	DA B	EB 1	Joe	44.76
11	DB 20	DA B	EB 1	Joe	43.12
12	DB 20	DA B	EB 1	Joe	44.66
13	DB 20	DA B	EB2	Joe	34.23

1 Production Run	2 Die Bake	3 Die Attach	4 Epoxy Batch	5 Operator	6 Die Shear
14	DB 20	DA B	EB2	Joe	34.13
15	DB 20	DA B	EB2	Joe	32.82
16	DB 20	DA B	EB2	Sue	33.11
17	DB 18	DA 1	EB 1	Sue	44.63
18	DB 18	DA 1	EB 1	Sue	44.78
19	DB 18	DA 1	EB 1	Sue	43.04
20	DB 18	DA 1	EB 1	Sue	43.41
21	DB 18	DA 1	EB2	Sue	33.19
22	DB 18	DA 1	EB2	Sue	32.39
23	DB 18	DA 1	EB2	Sue	34.69
24	DB 18	DA 1	EB2	Sue	35.41
25	DB 18	DA B	EB 1	Sue	44.66
26	DB 18	DA B	EB 1	Sue	44.36
27	DB 18	DA B	EB 1	Sue	43.53
28	DB 18	DA B	EB 1	Sue	44.33
29	DB 18	DA B	EB2	Sue	33.29
30	DB 18	DA B	EB2	Sue	33.90
31	DB 18	DA B	EB2	Sue	35.17
32	DB 18	DA B	EB2	Sue	32.90

The next step is to apply a partitioning analysis of variance technique (ANOVA) to the dataset in Table 19. We are attempting to create a mathematical model that predicts Die Shear as a function of the 4 variables (Die Bake, Die Attach, Epoxy Batch and Operator).

The ANOVA technique provides a measure, R-squared (R^2), as an indicator in percentage of how much of the observed Die Shear variation can be explained by the 4 variables of interest. In this case $R^2 = 97.1\%$. This means that 97% of the variation in Die Shear can be explained by the 4 variable mathematical model! A low value, for example $R^2 < 10\%$, would indicate that Die Shear is not explained by the 4 variables, which would suggest keep looking!

Since we have a high R^2, we can now partition the variation, using the Sum of Squares feature of ANOVA, to determine which variable(s) are primarily responsible for driving Die Shear. Your software should provide the option to display an ANOVA table much like the one in Table 20.

Table 20 ANOVA Table

Source		Sum of Squares	Percent Explained
Total Explained by Regression Model		844.439	
Variables	Die Bake	0.091	0.0%
	Die Attach	0.056	0.0%
	Epoxy Batch	843.749	97.0%
	Operator	0.543	0.1%
Unexplained Error		25.645	2.9%
Total Sum of Squares		870.084	97.1%

In this case, we found our special cause - it is the epoxy batch which explains 97% of the Die Shear Variation! So even though the control chart shows multiple OOC events, there is only 1 special cause - the epoxy batch. This technique is very effective at finding special causes when simple audit investigations do not reveal the root cause.

By the way, the Epoxy Batch special cause was traced back to 2 different suppliers of the epoxy. We found that epoxy specifications were different for each supplier!

In general, once you locate and correct the special cause:

- remove the data points associated with *Lurking Variable* from the control chart;
- conduct additional process runs with the corrective action in place to achieve the original (n_{init}) process run count;
- plot the new data points from post corrective action process runs in chronological order for all process runs and re-apply the *Lurking Variable* rules (2, 3 and 4);
- repeat until you pass.

I was always amazed how much an organization learns about their process and how much the process immediately improves when they successfully complete this step. This analysis always seemed to find the most obvious special cause(s) and when identified during setup was the easiest to quickly address.

7.12. Calculate the *Stability Index* (S_i) and Apply the 85% Rule

At this point you should have a control chart without any obvious *Lurking Variables* i.e., the chart passes rules 2, 3 and 4. It does not mean there are no special causes affecting your process. You may notice that I did not separately apply Rule 1 (1 Point out of Control). I say separately because, 1 point out of control can contribute to Rules 2, 3 or 4 violations. This is very intentional. Before you separately apply Rule 1, I want to be sure that the control chart reflects suitable random and independent samples, control limits are adjusted for (n_{eff}), *Lurking Variable* corrective actions have been applied, and replacement process runs have been completed, and the data has been re-plotted and re-analyzed.

Now I want you to calculate the ***Stability Index*** **(S_i).**

Stability Index **(S_i):** Count the number of control chart process runs that pass Rule 1 (1 Point out of Control) and divide by the total number of process runs plotted on the control chart. This fraction is defined as the *Stability Index* (S_i) and when converted to percentage reflects the % In-Control.

Finally apply the **85% Rule**.

85% Rule is defined as ≥ 85% of process runs are in-control when considering Rule 1. So, the control chart is considered *well-behaved* and the process is defined to be *reasonably stable*. It is calculated by converting the ***Stability Index*** **(S_i) ≥ .85** to a percentage (85%).

Consider Figure 18 that shows an example that fails the 85% Rule. The control chart indicates there are 6 OOC process runs failing Rule 1, there are 24 process runs in control (IC) per Rule 1 out of 30 total process runs being considered. This calculates out to be a *Stability Index* (S_i) of .80. When converted to a percentage, it indicates 80% of process runs are IC. This control chart fails the 85% Rule.

Figure 18 Example Fails 85% Rule (24/30 = 80%)

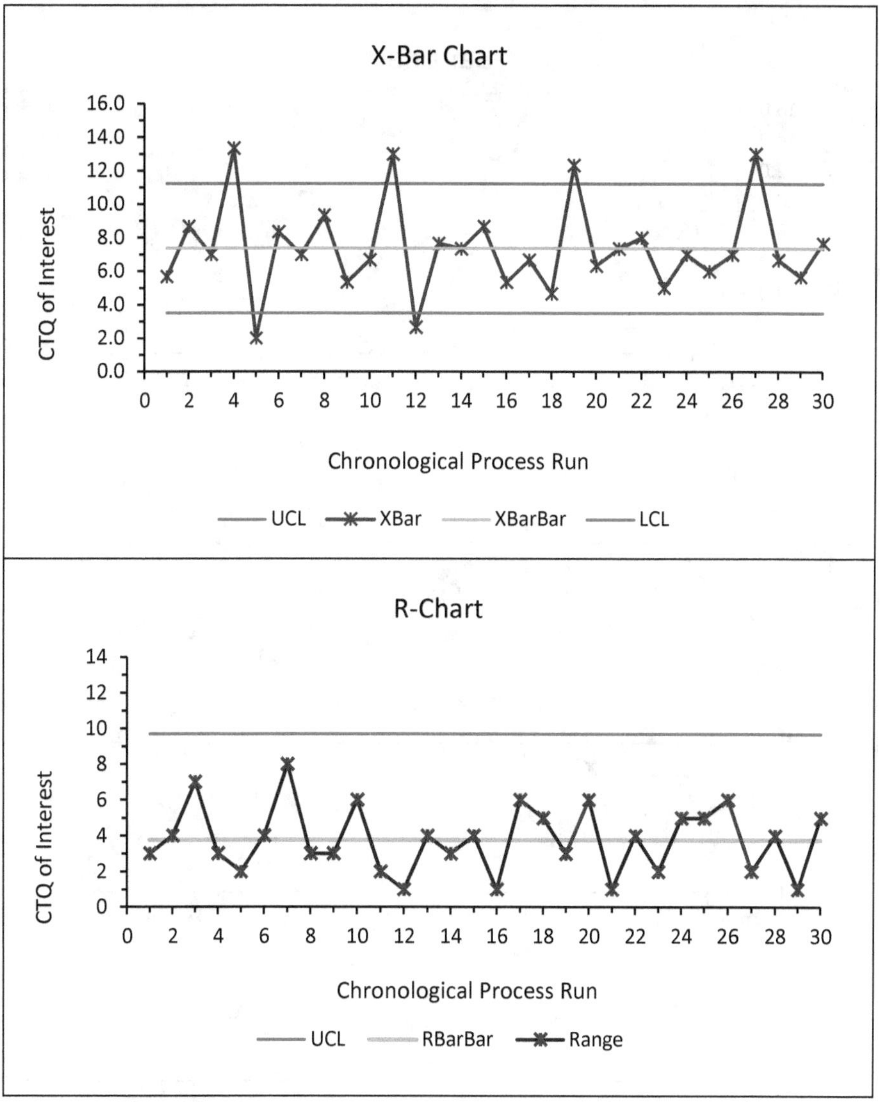

Consider Figure 19 where there are 3 OOC process runs failing Rule 1, 27 in control (IC) process runs out of 30 process runs, which calculates out to be *Stability Index* (S_i) = *.90* of process runs are IC. When converted to percentage, it indicates 90% of the process runs are IC. This example passes the 85% Rule but still indicates the presence of special cause(s), i.e., 3 Rule 1 violations.

Figure 19 Examples Passes 85% Rule (27/30 = 90%)

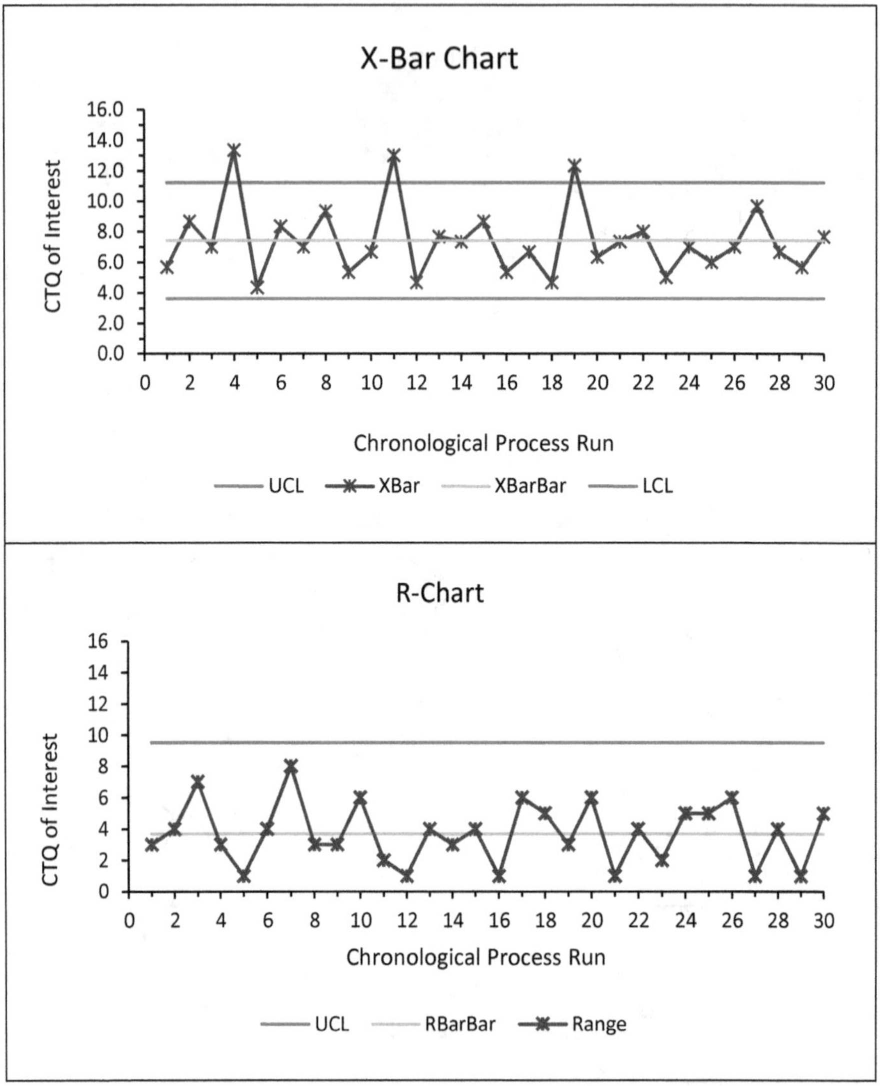

So why is the **85% Rule** so important and included as part of this control chart setup process? Expecting 100% in-control (IC) process runs during control chart setup is unrealistic even though we already took steps to address and remove *obvious lurking variables*. **Most importantly, this method double checks (n_{eff}).** My experience indicates that if you find < 85% of process runs are IC at this step, there is a good chance your (n_{eff}) calculation is in error. Go back to step 7.7 and be sure the analysis includes at least 30 process runs where the subgroup data from *Lurking Variable* OOC events were replaced with process runs that had *Lurking Variable*

corrective action(s) applied and plotted on the chart in chronological order.

There is another reason why the 85% Rule is so important. Classic statistical books define that estimates of product CTQ population parameters μ *(average)* and σ *(standard deviation)* are considered inaccurate unless the process is stable. This interpretation extends to Process Capability Indices C_p, C_{pk} (See Chapter 10) because you need valid μ and σ to calculate the indices. So, we are in a Catch-22 situation. You cannot calculate these population parameters for product CTQs or process capability indices when the process is not stable, yet everyone (notably customers) always wants to know these indices for Product CTQs. To address this situation P_p, P_{pk} Process Performance Indices were developed but are also often misused.

It gets worse because most SPC production software routinely calculates these capability indices independent of whether the control chart indicates that the process is stable. Indeed, production software often allows process capability calculations for each process run! Again, I approached some of the major SPC software suppliers and asked whether they were aware of this contradiction. They again indicated that they were aware of the contradiction, but offered this feature based on customer demand. Some newer software introduced P_p, P_{pk} features to address this issue.

So, I invented this practical compromise solution: ***The 85% Rule***. When you apply and pass the 85% Rule *at this step* of the control chart setup process, I am *defining* the control chart to be ***Well-Behaved*** and the process to be ***Reasonably Stable*** - and most importantly the control chart is almost **ready to be released to production**.

You can now use the process run dataset to calculate μ and σ and therefore the C_p, C_{pk} process capability indices for Product CTQs. (Recall Process CTQs should not have specifications so there are no capability indices).

Passing the 85% Rule means the control chart is *well-behaved*, the process is *reasonably stable*, and the control chart is *almost* ready to be released to production. The population parameters can now be used to calculate product CTQ capability indices.

So, what remains? Selecting and turning on OOC rules and writing OCAPs!

7.13. Define Quick Response OCAPs

Now it is time to define and document "Quick Response" Out of Control Action Plans (OCAPs). I purposely use the term "Quick Response OCAP" because the OCAP should be *initiated* immediately upon notification of an OOC event (i.e., does not require the presence of process engineering). Modern SPC Software can be set up to immediately notify an OOC event once the CTQ data has been entered into the system. Some software also allows the process engineer to define OCAPs that pop-up on the computer where the data has been entered, yet most OCAPs are defined in the process's documentation.

The best "Quick Response" OCAPs consists of 5 steps as described in Table 21. Formal Checklists are recommended to ensure completeness, consistency, and the ability to compare documented investigation results with other OOC events.

Table 21 OCAP Steps

OCAP Steps	Actions
1	• Production Personnel immediate investigates per Checklist 1.
2	• If special cause not found, Process Engineering investigates per Checklist 2.
3	• If special cause still not found, Process Engineering performs partitioning ANOVA analysis on the dataset.
4	• Process Engineering Removes Special Cause.
5	• Carefully monitor the next 8 process runs to ensure the process returned to its historical performance.

Step 1 should be those activities that *production personnel* can complete while waiting for process engineering to arrive as they should have been concurrently notified. Production Personnel should complete the investigation using a pre-defined checklist documented in approved work instructions. Table 22 provides examples of Investigation Categories (Column 1) and more detailed Areas to Address (Column 2) recommendations that can be used to create an OCAP checklist for production personnel. It is not intended to be a complete list. So be sure to include other activities judged pertinent to your situation. The investigation categories are derived from my experience for organizations that are in the initial stages of SPC deployment. They represent the most frequent sources of special causes for early SPC implementors. As the organization learns, the checklist is usually and should be updated. For example, once an organization has a well-established MSA program, investigating measurement error can be replaced with more pertinent actions.

Table 22 Production Personnel Investigative Activities

Investigation Category	**Areas to Address**
• Measurement Error	• Was a new Measurement Instrument used? • Was Measurement Instrument calibration up to date? • Remeasure samples and verify results.
• Equipment Repair or Maintenance	• Is maintenance up to date on the equipment used for the OOC event(s)? • Was recent repair or maintenance performed on the equipment used for the OOC event(s)?
• Documentation	• Was there a newly released process document? • Was the process parameters/execution changed? • Were all production personnel suitably trained to the newly release process documentation?
• Process Parameters	• Verify process run(s) for the OOC event(s) were run according to approved documentation.
• Adjustments	• Was an authorized process change used for the OOC event(s)? • Was an unauthorized process change used for the OOC event(s)?

Production Personnel should provide the results of their Step 1 investigation to process engineering upon their arrival.

If Step 1 was unable to detect the special cause, process engineering would continue the investigation using the Step 2 checklist. Process engineering can use suggestions from Table 23 to build a Step 2 checklist. Again, it is not intended to be complete list and be sure to include other activities that are judged pertinent to your situation; and/or update as the organization learns.

Table 23 Process Engineering Investigative Activities

Investigation Category	Areas to Address
• Measurement Error	• If a new Measurement Instrument was used, was there adequate analysis (Calibration or MSA Studies) to ensure equivalency with other instruments before being released to production?
• Equipment Repair or Maintenance	• If recent repair or maintenance was conducted on the equipment used to run the OOC event(s), were their adequate test process runs prior to being released to production? • Are the test process run procedures documented and approved?
• Newly Released Equivalent Equipment	• If the equipment used for the OOC event(s) was on newly released equivalent equipment, verify the adequacy of analysis assuring the new piece of equipment can be used on the control chart (see Chapter 8).
• Operator(s)	• Verify Operator(s) measuring the CTQs for the OOC event(s) were certified to run the Measurement Instrument. • Verify Operator(s) performing OOC event(s) were certified to run the Process.
• Process Parameters	• Verify that the process parameter settings for the OOC event(s) were set to those specified in the most recently approved process document.
• Documentation	• If the OOC events were subject to a newly released process work instruction, verify adequacy of test runs prior to the release to production.
• Process Parameters	• If an authorized process adjustment(s) were used on OOC event runs, verify adequacy of test runs prior to the release to production. • Was the process adjustment caused by an OOC event or an OOS event? • If unauthorized process adjustment(s) were used on OOC event runs, investigate why the process was allowed to run with an unauthorized process change.

For the situations where Step 2 did not reveal the special cause, the more rigorous Step 3 analysis may be necessary. The most successful approach I found was to use the same partitioning ANOVA method described in 7.11 where I discussed how to locate *Lurking Variables*. First, you will need to build a response table like Table 19 Die Attach Audit Results. Often the results from the Quick Response checklists audits can be used to help build the response table. Once analyzed, the R^2 values from the ANOVA table (Example: Table 20) are used to locate the special cause(s).

The next step, Step 4, is to *remove* the special cause. This means to take actions to eliminate or remove the special cause – not adjust for! Table 24 provides examples of common special causes and effective removal strategies.

Table 24 Examples of Special Cause Removal Strategies

Special Cause	Removal Actions
• Measurement Error caused by uncalibrated equipment	• Remove offending instrument from use until it is properly calibrated and found to be equivalent to other qualified measurement equipment. • Update policies and procedures to ensure only up-to-date calibrated measurement equipment can be used in production.
• Repair Event	• Remove equipment from use until re-qualified. • Update policies and procedures to ensure repaired equipment is re-qualified as in control prior to releasing back to production. • Release back to production once re-qualified.
• Authorized Adjustment	• Remove equipment from use until re-qualified. • Update policies and procedures to ensure adjusted equipment is re-qualified as in control prior to use in production. • Release back to production once re-qualified.
• Newly Released Equivalent Equipment	• Control the newly released equivalent equipment on its own separate control chart until it is found to be statistically equivalent (see Chapter 8).

Special Cause	Removal Actions
• Single Operator	• Re-train, re-certify operator until they demonstrate ability to run the process in control. • Audit their process execution for 5 or more runs.

The final action, Step 5, is to carefully monitor the next 8 process runs to ensure that after special cause removal, that the process has returned to its historical performance. In other words, if the next 8 process runs plotted on the control chart remain in control, your corrective action worked! I suggest carefully monitoring 8 process runs since it allows you to apply all the relevant special cause rules.

These Quick Response OCAP recommendations are based on my experience observing which investigative actions were the most effective and quickest at determining the special cause(s) responsible for the OOC event(s). This 5 Step Quick Response checklist approach located the special cause about 80% of time within 2 days, often within 24 hours. Further, documented checklists and their results enabled the organization to "learn" more about their processes and procedures and often resulted in checklist updates.

The **least effective OCAPs were "Call Engineering" OCAPs** that allow continuous operation of the process until Process Engineering arrives and decides to do something. Invariably the process continues with an active special cause until the process engineer arrives. This often took days. "Call Engineering" OCAPs often allow Process Engineering to authorize a process adjustment, rather than complete an investigation. Be sure to re-read "The Good, The Bad and The Ugly OCAPs" discussed in section 5.5.

7.14. Define OSAPs for Product CTQs (when necessary)

Recall that CTQs may be Process or Product characteristics. Product CTQs typically have specifications. Customers expect Out of Specification Action Plans (OSAPs) will be implemented whenever a production unit is measured out-of-specification (see Chapter 4). These OSAPs should ensure that customers will not receive out-of-specification products. So, if any production unit within a subgroup sample is measured out-of-specification, the OSAP should be implemented independent of whether the process run was in or out of control. However, ***OSAPs actions should be limited to the***

containment and disposition of production units measured out of specification. For example, disposition actions can be to scrap or rework (if re-work is allowed) the OOS production units. OSAPs could also include additional assessment activities on other production units within the process run. Be sure to document OSAPs in your work instructions and clearly separate OCAPs and OSAPs.

However, OSAPs should never include process adjustments. Recall how competing priorities (Chapter 4) and Specification Limits on Control Charts (Chapter 5.2) are major contributors to *Ineffective Control Charts.*

If you continuously find sampled production units are out-of-specification, it is likely you have a bad specification. If this case read Chapter 11, where I not only discuss how to determine if your process can meet your specification, but I also discuss how to set Product CTQ specifications to meet process capability requirements.

7.15. Enable Out-of-Control (OOC) Rules with Quick Notifications

Now it is time to enable Out-Of-Control (OOC) Rules that include Quick Notifications that will trigger the OCAPs defined in 7.13. Figure 20 reminds us of the 4 most used control chart rules. The good news is if you are using modern production SPC software, you should have the option of independently enabling each rule (or others) for each control chart and setting up Quick Notifications.

Figure 20 Control Chart Rules

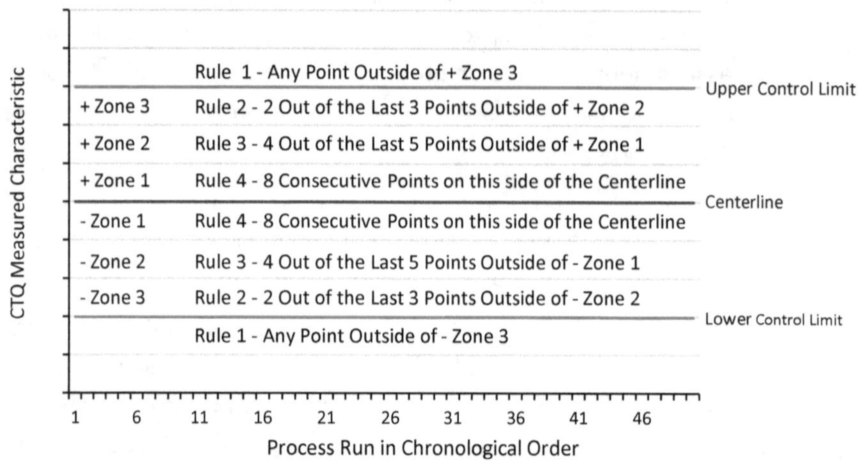

Figure 21 provides examples of special cause rules that can be applied to new process runs once the control chart was released to production. Since the recommended control charts (*X_i & MR, $\overline{\mathbf{X}}$ & R, or $\overline{\mathbf{X}}$ & s*) are all 2-parameter control charts where you are monitoring and plotting the CTQ's process average (μ) and variation (σ), the rules are applied independently to each CTQ's parameter control chart to determine violations. Figure 21 Examples of Special Cause Rule Violations also reminds us that the rules are applied symmetrically on both sides of the centerline.

Figure 21 Examples of Special Cause Rule Violations

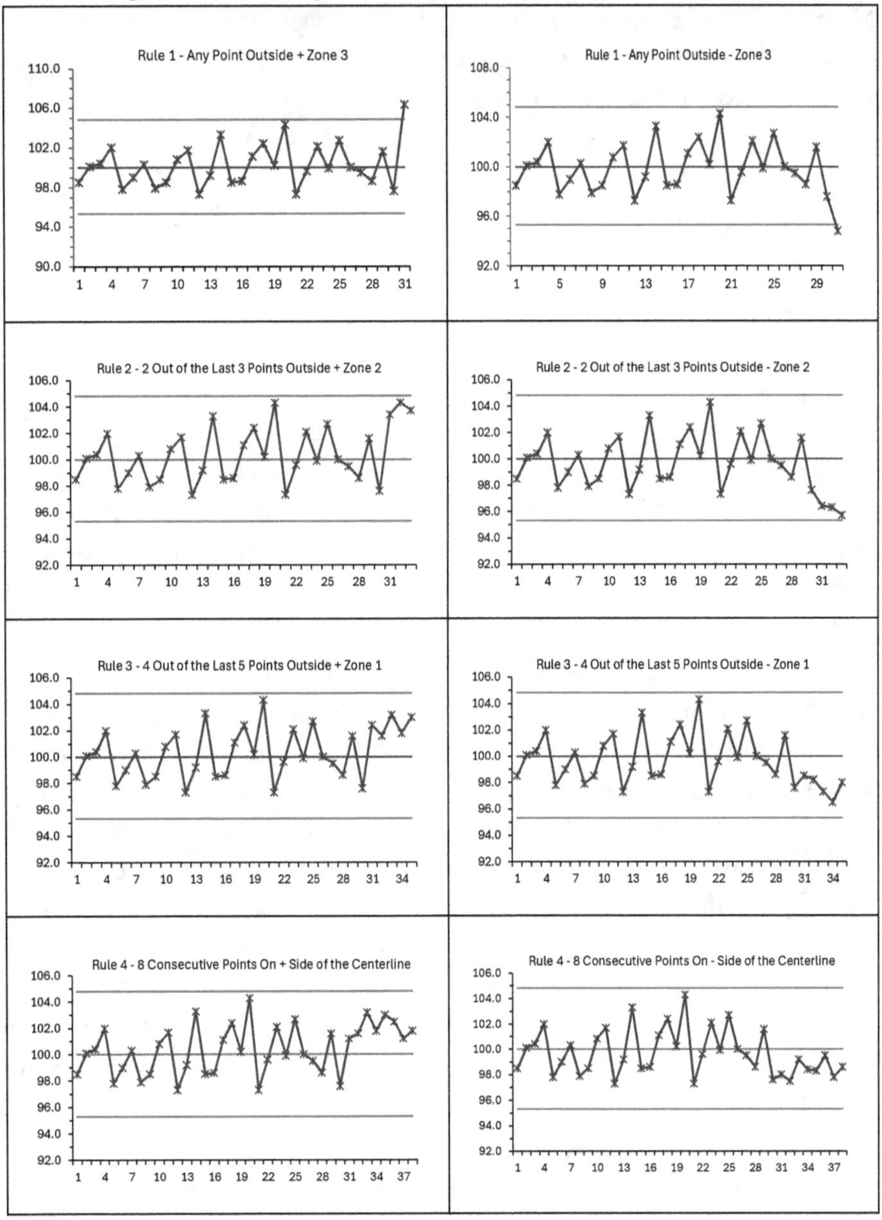

In these examples the first 30 runs were used to set up the control chart. If you follow my control chart setup guidelines, it means you *qualified* and released a *well-behaved* control chart monitoring a *reasonably stable process.* Therefore, the special cause rules are applied to process runs beginning with run 31 and beyond.

Modern SPC software allows you to turn these SPC rules on and off at will. Most SPC Chart owners turn them all on when first releasing a control chart - just as classic SPC training suggests. This is where my recommendations may differ from others. If you follow the methods described in this chapter, you only need to implement 2 Special Cause Rules after you release a *Qualified (as defined in this chapter)* control chart. The 2 rules are:

- Rule 1: One point-out-of-control.
- Rule 4: 8 consecutive points on either side of the centerline.

The reason I recommend only enabling Rule 1 and Rule 4 is because they both indicate very serious process shifts. From a practical point of view, the special causes that trigger these OOC events are also the easiest to find when there are well-written Quick Response OCAPs. This suggestion is particularly helpful for organizations where they are experiencing too many OOC events across many processes within a factory where there are not enough resources to concurrently execute Quick Response OCAPs. But before you implement this suggestion, be sure you complete 7.11 Analyze and Remove Obvious *Lurking Variables* during control chart setup.

If your OCAPs are quick and diligent you will find the optimum balance of OOC event process stoppage, while maintaining stable processes. It also prevents the spread of ineffective OCAPs strategies, such as process adjustments. Just ensure that production operators *immediately* enter the data into the SPC production software upon receiving the process run CTQ measurements *before the next process run*. Sometimes your systems can be set up where the measurement equipment automatically enters the data into the SPC software. Also ensure the SPC software *immediately* notifies the operator and other defined personnel if an OOC event is triggered. I have seen many occasions when data collected by operators was not entered into the system until the end of the shift - often after many other process runs were completed - this practice often misses the initial OOC event!

Quick Notifications may also include additional actions which should be defined in the OCAP. For example, early in my career when I was a Plasma Etch Process Engineer, once an OOC event occurred, my notifications included setting a software flag on the process equipment that prevented further processing. I had to override the flag to resume processing. Worked every time and immediately!

I have seen many cases of ineffective control charts caused by delayed actions even after properly setting up the control chart per my recommendations. For example, the production software was not set up to notify production personnel upon detection of an OOC in real time. The OOC Rules were being applied and interpreted by process engineering just before a Weekly SPC Review meeting, where OOC events and possible actions would be discussed. In the meantime, the special cause that triggered the OOC and the process shift remained active. Clear recipe for ineffective control charts!

When an organization experiences frequent OOC events, they often resist quick response. They often resort to authorizing adjustments and delay investigative actions until SPC review meetings (often scheduled weekly) to make decisions on which OOC event(s) to investigate! Think about it, a detectable OOC event occurred on a well-behaved control chart which indicates that the process has shifted, i.e., a special cause has introduced additional variation to your process, and you ignore it until the SPC meeting! What happens if an investigation is not authorized? The special cause that just added variation to your process continues.

This often results in many OOC events, causing many OCAPs investigations and many process stoppages. Because management does not want to stop processing, OCAPs begin changing from stopping the process to investigate the special cause to the other less effective OCAP actions, such as just making process adjustments.

So, my suggested approach is a compromise that has proven to be the most effective at keeping control charts effective. When Rule 1 (one point outside the UCL or LCL) is violated, the special cause is most often found with a quick and simple audit of OOC process run. When Rule 4 (8 Consecutive Points on One Side (+/-) of the Centerline) is violated, it indicates the process average has shifted and is still present. The special cause is most often the result of an authorized or unauthorized process adjustment. Even if it is not obvious, the root cause is most often found using the previously discussed Analysis of Variance (ANOVA) technique.

It is not that the other rules will not work. They do work, yet frequent OOC events interfere with Quick Response and cause the deterioration of OCAP effectiveness - competing priorities sneak in - rendering the control chart ineffective.

7.16. Release the Qualified Control Chart to Production

Congratulations when you reach this step! This means that you have developed a *Well-Behaved Control Chart* for a *Reasonably Stable* process. The control chart is now ready to be released to production and is now defined by me to be ***Qualified Control Chart.***

RECAP

Table 25 recaps the qualification setup requirements for a Control Chart as well as key issues to address for each step.

Table 25 Qualification Requirements for a Control Chart

Qualification Requirements	**Key Issues/Comments**
1. Identify the Process, Process Equipment, and the CTQ.	• Can be a Process or Product CTQ • Be aware and address "Copy Exact" or "Copy Similar" equipment.
2. Verify the Readiness of Measurement Equipment.	• Minimum requirement is completed Gage R&R studies. • Comprehensive MSA studies should be completed if multiple Measurement Systems are used or needed.
3. Verify Work Instructions and Production Operators are Certified.	• Documentation should be formally released and approved for both process and measurement system documentation. • Ensure Production Operators are certified to the *latest revision!*
4. Define the Initial Sampling Plan (n_{init}).	• More sampling is better because Effective Sample Size (n_{eff}) analysis will be more accurate. • Used for the initial 30 process runs. • Be sure to consider known equipment features and anticipated sources of variation.
5. Setup Trial Control Limits for Pilot Runs	• Use C_{pk} formula to calculate trial control limits with a value between 1 and 1.33. • Use specification limits if it is a Product CTQ. • Use Maximum and Minimum accepted values for Process CTQs.

Qualification Requirements	**Key Issues/Comments**
6. Conduct 30 Process Runs and Measure the CTQ.	• Only certified production personnel are allowed to conduct the process runs. • Follow only approved work instructions. • Process Changes require re-starting the 30 process runs.
7. Analyze the Dataset for the Effective Sample Size (n_{eff}).	• Usually (n_{init}) ≠ (n_{eff}). • Calculate the effective sample size using Equation 4.
8. Determine the Most Random and Independent Samples.	• Use R^2 Correlation Matrix/Coefficient to select the most uncorrelated site(s). • Observe/Audit the process to determine a physical meaning for (n_{eff}).
9. Decide if the Ongoing Sampling plan Should be Reduced to (n_{eff}).	• Okay to keep the (n_{init}) sampling plan temporarily or permanently but be sure to use (n_{eff}) for control limit calculations.
10. Construct a Control Chart using (n_{eff}) to set Control Limits.	• $\bar{X}$ & s is best when (n_{eff}) >2. • $\bar{X}$ & R is best when (n_{eff}) = 2. • X_i & MR is best when (n_{eff}) = 1. • Others work but are not needed and are even harder to properly setup.
11. Analyze and Remove Obvious *Lurking Variables*.	• Most prevalent lurking variable: equivalent "Copy Similar" equipment combined too early on the control chart.
12. Calculate the *Stability Index* (S_i) and Apply the 85% Rule.	• Ensures a well-behaved chart monitors a reasonably stable process. • Allows re-checking (n_{eff}). • Enables C_p, C_{pk} process capability indices calculations.
13. Define Quick Response OCAPs.	• Be sure investigations start immediately upon an OOC event. • Use checklists. • Update OCAPs as the organization learns. • Do not use process adjustments as an OCAP.

Qualification Requirements	Key Issues/Comments
14. Define OSAPs for Product CTQs (when necessary).	• OSAPs actions should be limited to the containment and disposition of production units measured out of specification. • Do not use process adjustments as an OSAP.
15. Enable Out-of-Control (OOC) Rules with Quick Notifications.	• Best balanced suggestion: You only need to enable Rules 1 & 4!
16. Release the Qualified Control Chart to Production.	• Congratulations!

8. COMBINING EQUIPMENT ON CONTROL CHARTS

Statistical Methods include methods to determine equivalence between samples. For example, you can use the *t-test* to compare averages and the *F-test* to compare variances of samples to see if they are significantly different. If the results indicate that the two populations are different, you have a definitive answer.

Control charts are more valuable for ongoing process monitoring and control as they help understand process behavior over time and identify any unusual patterns or shifts (special causes). Which is better? For comparing groups or variances at a specific time or dataset, use the *t-test* and *F-test*. For ongoing process stability and special cause variation, use Control Charts with Special Cause Rules.

Most production facilities have multiple pieces of equivalent equipment intended to deliver the same process, and sometimes more are added to increase capacity. Statistically equivalent equipment means that two or more sets of data from equivalent equipment are considered similar enough that they can be treated as belonging to the same process. In other words, their statistical properties, such as mean and variability, do not show significant differences, which suggests that the processes producing these data sets are behaving in a consistent and predictable manner. The objective is to combine data from statistically equivalent equipment on one control chart while maintaining separate charts for non-statistically equivalent equipment, as combining statistically equivalent process equipment on the same chart can be convenient.

Non-statistically equivalent equipment can still deliver similar process outcome. However, combining them on the same control chart without verifying statistical equivalence leads to false OOC events. As previously discussed in Chapters 4 and 5, false OOC events trigger a "*domino effect*" that results in ineffective SPC charts.

As noted in Chapter 5, equipment may be "Copy Exact" or "Copy Similar." "Copy Exact" equipment comprises identical units with the same work instructions and processing parameters. "Copy Similar" equipment includes identical, upgraded, or alternative models with equipment specific processing parameters, all aiming for the same process outcome.

How do you combine multiple pieces of equivalent equipment on the same control chart? Note that these methods also apply when the CTQ is measured after a sequence of processing events such as the example described in 7.11.

This approach assumes you have a previously qualified SPC chart using the methods described in Chapter 7.

8.1. Combining "Copy Exact" Equipment

First, confirm that the work instructions and ***processing parameters match exactly***. If they do not, treat the case as "Copy Similar" and refer to Section 8.2. If confirmed, you can use the same *Effective Sample Size* (n_{eff}) and sampling plan as the Qualified chart or repeat the qualification steps in Chapter 7 to verify (n_{eff}). If you use the same *Effective Sample Size* (n_{eff}), conduct *at least* 8 process runs using the candidate equipment and plot the points on the Qualified Chart. Eight Process runs allow you to apply all Special Cause Rules 1-4.

Figure 22 displays a combined control chart. The initial 30 runs are from the primary equipment's qualified chart, while runs #31-40, plotted in chronological order, are from the candidate equipment. Sampling was conducted based on *Effective Sample Size* (n_{eff}) as detailed in Chapter 7. No out-of-control (OOC) events were observed after applying Special Cause Rules 1-4.

Figure 22 Data Includes "Copy Exact" Process Runs

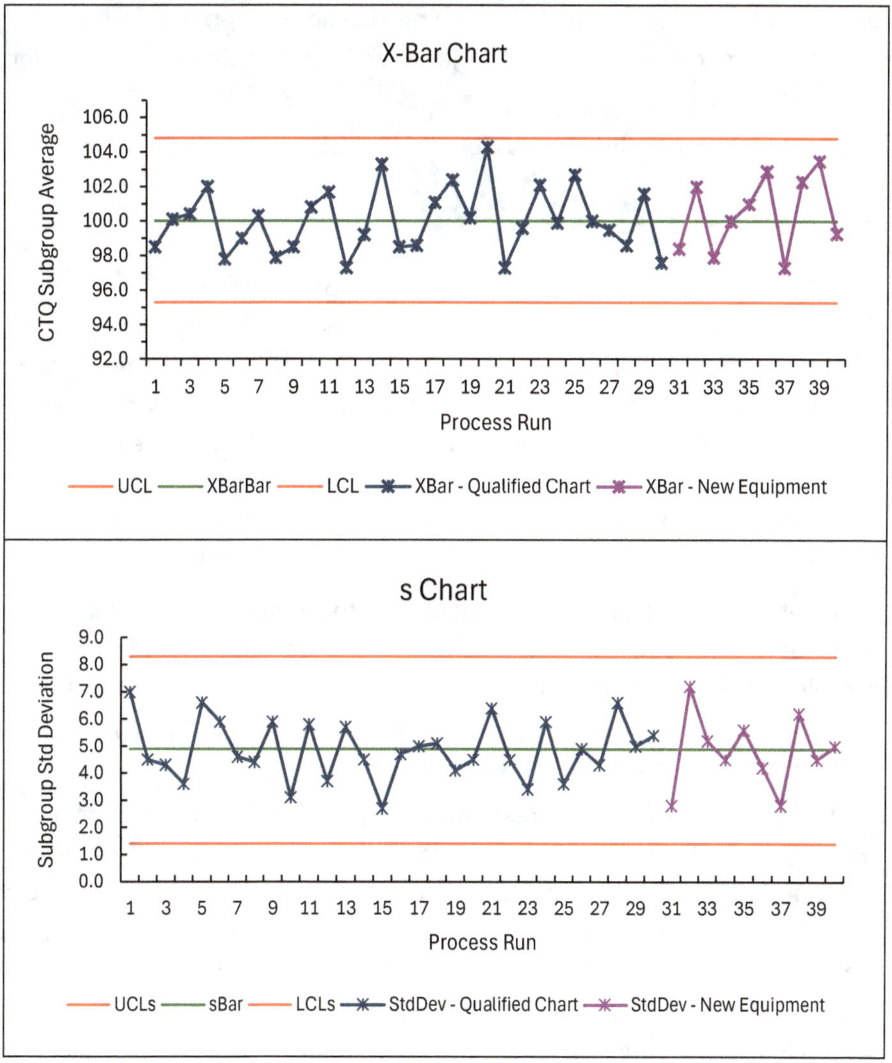

If you achieve these results, congratulations! You can now combine data from equivalent equipment on the same control chart and start plotting all data chronologically.

If a Special Cause rule is violated, it indicates that the "Copy Exact" equipment is not providing statistically equivalent results. In this situation, you have two options. Option 1 is to investigate the "special cause." Once identified and resolved, conduct at least another 8 process runs, and re-test. Once the Special Cause Rules pass, the equipment can be combined on the same chart. Option 2 is to maintain the "Copy Exact" equipment on its own separate control chart. This method is effective when both pieces of equipment produce products within specifications. Keep in mind, modern SPC software can easily be configured to monitor different equipment separately.

Be sure to repeat this process for all new "Copy Exact" equipment.

8.2. Combining "Copy Similar" Equipment

This situation is common and challenging. CTQ outputs from "Copy Similar" equipment should be statistically equivalent, but they often are not initially. Instead of waiting for statistical equivalence, some Process Engineers combine these on the same control chart if a few samples fall within control limits. This issue is worse with Product CTQs, where equipment is combined on the same chart if samples meet specifications. This improper practice can cause false Out-of-Control events and render the control chart ineffective.

What complicates the issue further is that the *Effective Sample* (n_{eff}) might differ between "Copy Similar" equipment. As we know, using the wrong sample size during control chart setup often results in incorrect control limits, usually too tight, causing false OOC events. These false OOCs affect OCAPs and can render the control chart ineffective.

The best approach is to treat each "Copy Similar" equipment as a new process and follow all the procedures described in Chapter 7 on how to setup and qualify a control chart. Once this is done you have established that the "Copy Similar" process equipment is *well-behaved* and *reasonably stable,* and you have a qualified control chart. You can then plot the same process runs on the previously qualified chart as described in 8.1. If you pass the Special Cause Rules 1-4, then you can combine equivalent equipment on the same control chart. Often the calculated control limits on the "Copy Similar" chart are so close to the already established equipment, the decision is obvious!

If you fail the Special Cause Rules, you have the same options as described in 8.1. Option 1 is to investigate the "special cause." Once identified and resolved, conduct at least another 8 process runs, and re-plot. Once the Special Cause Rules pass, the equipment can be combined on the same chart. Option 2 is to maintain the "Copy Exact" equipment on its own separate control chart. This method is effective when both pieces of equipment produce products within specifications. Keep in mind, modern SPC software can easily be configured to monitor different equipment separately.

While having equipment on their own control chart may seem more complicated, modern SPC software simplifies the process. The goals of SPC are to ensure that process equipment produces consistent results and to reduce process variability by investigating and addressing special causes. Keeping statistically non-equivalent equipment on separate control charts can decrease the organization's workload. This approach allows for stopping the specific piece of equipment involved in an OOC event for investigation without disrupting the operation of other equipment. Combining equipment before proving their equivalence leads to numerous issues, such as competing priorities and ineffective OCAPs (adjustments, engineering calls, etc.). This situation significantly increases workload as the organization deals with false OOC events, ultimately resulting in ineffective control charts.

9. WHEN TO RECALCULATE CONTROL LIMITS

Recalculating control limits on a control chart is a crucial element of maintaining an effective SPC Program as well as capturing the promise of reducing process variability. When a special cause has been identified and removed, recalculating the control limits is essential to ensure process variation improves. Applying tighter control limits on the process as special causes are addressed is necessary to achieve the process improvement potential of Control Charts. Keeping the control limits unchanged will only maintain the process's historical performance.

Here are some situations when you should consider recalculating the control limits:

1. After Identifying and Removing a Special Cause.
2. After a Deliberate Process Adjustment.
3. After Specified Review Periods for Well-Behaved Processes.

9.1. After Identifying and Removing a Special Cause

If you follow this book's suggestions, you apply only Rules 1 and 4 (Section 7.15) after releasing a *Qualified Control Chart*. Using this approach, most out-of-control events will appear on the sample average chart rather than the variation chart and are easy to find. After investigating and applying corrective actions to remove the special cause, conducting 10 process runs on the equipment which had the violation and re-apply Rule 4 without recalculating the control limits. If no further violations are detected it indicates your actions were effective, perform an additional 20 more runs on the equipment under consideration for a total of 30 process runs since the special cause removal, and recalculate the control limits. You will likely see reduced control limits. Reduced control limits indicate process improvement. If you continue to violate Rules 1 or 4 during the first 10 runs post corrective action, it means your investigation and removal actions did not work.

9.2. After a Deliberate Process Adjustment

Deliberate process adjustments are often made when there is a need to shift the process average to enhance product performance. It is less common to use deliberate adjustments to target and improve process variation performance, as this is generally more challenging. Intentional process adjustments are also used to move the process average and align non-equivalent equipment with preferred equipment. This allows for combining control charts and improves overall process variability as process equipment becomes statistically equivalent.

After making the process adjustment, run the process 10 times and apply Rule 4. If you detect a violation, you have successfully shifted the process average. If the new average meets your goal, perform 20 more runs with the equipment and recalculate control limits based using the last 30 process runs which included the process adjustment.

9.3. After Specified Time Periods for Well-Behaved Processes

Recalculating control limits after Specified Time Periods is also essential for maintaining accurate and effective control charts. Here are some reasons why it is important:

- *Improving Accuracy:* As more data is collected, recalculating control limits can improve their accuracy. Initial control limits may be based on a limited dataset (30 data points), but with more data, the estimates of the process average and variation become more precise.
- *Reflecting Current Process Performance:* Over time, processes can undergo subtle changes, even if no deliberate adjustments are made. Regularly recalculating control limits ensure they accurately reflect the current state of the process and its inherent variation.
- *Detecting Changes in Process Behavior:* Regular recalculations help identify shifts or trends in process behavior that may not be immediately apparent. This allows for timely interventions to address any emerging issues before they become significant problems.
- *Ensuring Compliance:* In regulated industries or to ensure compliance with Quality Standards or customer requirements, there may be requirements to periodically review and update control limits to demonstrate ongoing process control and compliance.

Specifying the Review Time Period for recalculating control limits on a control chart on a regular basis can be challenging. It depends on several factors, but most importantly process stability and frequency of process runs.

Please recall the 85% Rule as described in Section 7.12, which indicates that the qualified control chart released to production represents a process that is well-behaved and reasonably stable. My recommendation is that Specified Review Period for control chart recalculations should be conducted only for those processes where the current control chart consists of at least 100 runs and continues to meet the 85% Rule. If the control chart fails to meet the 85% Rule, it is more prudent to investigate special causes rather than recalculating the control limits. Recalculating control limits when special causes are evident may result in the expansion of control limits, indicating either a shift in process performance or degradation in process variability.

The frequency of process runs also influences the Specified Review Periods. For example, a process that is run once a day or once a week generates much less data than a high-volume production facility that runs the process 15 times a day. Table 26 provides examples of process run frequency and suggested Specified Time Period.

Table 26 Examples of Specified Time Period

Process Runs	Process Runs Per Month	Process Runs Per Quarter	Suggested Specified Time Period
15 per day	~450	~1350	Once a Month
1 per day	~30	~90	Once a Quarter
1 per week	~4	~12	Once every 100 Process Runs

Ultimately, the best time period for recalculating control limits will depend on your specific process and operational context. It is important to strike a balance between having up-to-date control limits and not overburdening the organization with constant recalculations. By recalculating control limits at specified intervals, you can ensure that your control charts remain a reliable tool for monitoring and maintaining process stability and quality.

10. CAPABILTY AND PERFORMANCE INDICES

The primary reason for this Chapter is to introduce and review the C_{pk} and P_{pk} Indices to set up Chapter 11 on how to use these indices to calculate more realistic product specifications. So, in this chapter I want to cover a couple of other topics: a basic overview of these indices; and the most common errors when using these indices.

The concept of process capability and performance indices have their roots in the early quality management and statistical process control (SPC) movements, which emerged in the mid-20th century. Pioneers like Walter A. Shewhart, W. Edwards Deming, and Joseph M. Juran made significant contributions to the development of statistical methods for quality control. These methods laid the foundation for modern process capability indices. The exact historical timeline for the development and definition of C_{pk} (Capability Index) and P_{pk} (Process Performance Index) is not well-documented. However, these indices are rooted in the broader field of statistical process control (SPC) and quality management, which gained prominence in the mid-20th century. The development of process capability indices played a crucial role in advancing quality management practices, enabling organizations to consistently produce high-quality products and improve their processes.

As previously discussed, Critical-To-Quality (CTQs) characteristics monitored by a control chart may be Process CTQs or Product CTQs. Process CTQs should not have specifications, while Product CTQs should have specifications. Production's goal is to continuously deliver products that meet specifications. These indices were invented to provide a quantitative measure of how well a manufacturing or business process can produce products that meet CTQ specifications.

Key Reasons why Process Capability and Performance Indices are important:

- By using universally accepted indices like C_p, C_{pk}, P_p, and P_{pk}, organizations can consistently evaluate and compare the capability of different processes, both within the organization, across various equipment, suppliers, and across industries.
- Process capability and performance indices provide objective, data-driven information that helps management make informed decisions

about process control, quality improvements, and resource allocation.

- Many industries have stringent quality standards and regulations including Process Capability requirements. Process capability indices help organizations demonstrate compliance with these standards, ensuring that their processes meet the required specifications.

The most used process capability indices used for Product CTQs (Critical-to-Quality) are C_p, C_{pk} for stable processes. P_p, and P_{pk}, referred to as Process Performance Indices, were also introduced because they do not require stable processes to calculate. Both these indices help evaluate how well a process performs in relation to its desired CTQ specifications. A higher capability index value indicates the process is more likely to consistently meet specifications, while a lower capability index suggests the process may struggle to consistently meet specifications.

10.1. Review Formulas for Process Capability & Performance Indices

C_p PROCESS CAPABILITY INDEX

C_p Index measures the potential capability of a stable process by comparing the spread of the process variation (σ) to the width of the specification limits using the following formulas:

- Formula for Double Sided Specification Limits:

$$C_p = \frac{USL - LSL}{6\sigma_{within}}$$

- Formula for Single Sided Upper Specification Limit (USL):

$$C_p = \frac{USL - \mu}{3\sigma_{within}}$$

- Formula for Single Sided Lower Specification Limit (LSL):

$$C_p = \frac{\mu - LSL}{3\sigma_{within}}$$

Where: USL = Upper Specification Limit
LSL = Lower Specification Limit
u = Average of all Measurements.
σ_{within} = Standard Deviation of all within Subgroup Data

A *Key Assumption* for the C_p index introduced in this book is that C_p should be calculated using data from a stable and *well-behaved* process control chart. My suggestion is to use the *85% Rule* (a *Stability Index* ≥ .85) as described in Section 7.11 to determine control chart stability. The sigma (σ) used in the calculation of the C_p capability index represents only the within subgroup (short-term) variation. The C_p index focuses on the inherent process variation without considering any potential variation between different groups or shifts in the process over time. A higher C_p value indicates a more capable process.

A C_p value of 1.67 or higher is considered acceptable for most industries including Automotive, Aerospace, Pharmaceutical and Medical Devices for Product CTQs, meaning the process spread (variation) is well within specification limits.

C_{pk} PROCESS CAPABILITY INDEX

C_{pk} Index measures short-term process capability, considering within-group (short-term) variation and process centering, under the assumption that the process is stable and in control. The sigma (σ) variation used in the calculation of the C_{pk} is calculated from the within-group (short-term) variation, like C_p. It is the minimum value of the distance between the process mean and the specification limits, divided by three times the process standard deviation for a stable process using the following equations:

- Formula for Double Sided Specification Limits:

$$C_{pk} = \min\left(\frac{USL-\mu}{3\sigma_{within}}, \frac{\mu-LSL}{3\sigma_{within}}\right)$$

- Formula for Single Sided Upper Specification (USL):.

$$C_{pk} = \left(\frac{USL-\mu}{3\sigma_{within}}\right)$$

- Formula for Single Sided Lower Spec Limit (LSL):

$$C_{pk} = \left(\frac{\mu-LSL}{3\sigma_{within}}\right)$$

Where: USL = Upper Specification Limit
LSL = Lower Specification Limit
u = Average of all Measurements
σ_{within} = Standard Deviation of all within Subgroup Data
min = selecting the minimum of the 2 calculations

C_{pk} Index has the same *Key* Assumption as C_p, it should be calculated using data from a stable, well-behaved process control chart (*Stability Index*

$\geq$ .85). C_{pk} also only considers short-term variation only, so σ is calculated only from within subgroup dataset. C_{pk} focuses on a process's ability to produce the CTQ within specified limits, considering both the process's variation and whether the process average is centered between the limits. You can and should only include data from combined equivalent equipment that have been qualified to be joined on the same control chart when calculating C_{pk}.

A C_{pk} value of 1.67 or higher is considered acceptable to most industries including Automotive, Aerospace, Pharmaceutical, and Medical Devices for Product CTQs. It indicates that the process variability is well within the acceptable range, and the process average is centered close to the target value (center of specification).

P_p PROCESS PERFORMANCE INDEX

P_p Index measures the overall performance of a process by comparing the total process variability to the specification limits over a specified time period. It provides insight into how well a process can produce output that meets specified limits, considering both short-term (within subgroup variation) and long-term variation (between subgroup variation). P_p use the following formulas:

- Formula for Double Sided Specification Limits:

$$P_p = \frac{USL - LSL}{6\sigma_{overall}}$$

- Formula for Single Sided Upper Specification Limit (USL):

$$P_p = \frac{USL - \mu}{3\sigma_{overall}}$$

- Formula for Single Sided Lower Specification Limit (LSL):

$$P_p = \frac{\mu - LSL}{3\sigma_{overall}}$$

Where: USL = Upper Specification Limit
LSL = Lower Specification Limit
u = Average of all Measurements.
$\sigma_{overall}$ = Standard Deviation of all Measurements

A *Key Assumption* for the P_p index is that it does not require a stable and/or well-behaved process. So, the *Stability Index* is not needed to calculate P_p. It is acceptable to include all equivalent equipment process performance data from all control charts, i.e., the control charts are not required to be qualified, and all can be combined. They just need to reflect the same process and the same Product CTQ. P_p evaluates the overall (long-term) variability of the process, including both within subgroup and between subgroup variation without requiring the process to be in a state of statistical control.

P_{pk} PROCESS PERFORMANCE INDEX

P_{pk} Index measures the overall process performance, considering both short-term and long-term variations, and how well the process average is centered between specification limits. It is the minimum value of the distance between the process mean and the specification limits, divided by three times the overall standard deviation using the following equations:

- Formula for Double Sided Specification Limits:

$$P_{pk} = \min\left(\frac{USL-\mu}{3\sigma_{overall}}, \frac{\mu-LSL}{3\sigma_{overall}}\right)$$

- Formula for Single Sided Upper Specification (USL):

$$P_{pk} = \left(\frac{USL-\mu}{3\sigma_{overall}}\right)$$

- Formula for Single Sided Lower Spec Limit (LSL):

$$P_{pk} = \left(\frac{\mu-LSL}{3\sigma_{overall}}\right)$$

Where: USL = Upper Specification Limit
LSL = Lower Specification Limit
u = Average of all Measurements
$\sigma_{overall}$ = Standard Deviation of all Measurements
min = selecting the minimum of the 2 calculations

P_{pk} Index has the same *Key* Assumption P_p, it does not require a stable and/or well-behaved process. So, the *Stability Index* is not needed to calculate P_{pk}. P_{pk} data collected should be representative of all process equipment, capturing both short-term (within-group) and long-term (between-group) variations. So, it is acceptable to include all equivalent equipment process performance data from all control charts independent of whether they are qualified to be combined. They just need to deliver the same process and Product CTQ.

C_{pk} IS THE PREFERRED INDEX

Which index is most preferred by customers and industries? Evidence indicates that C_{pk} is commonly preferred by many industries for Product CTQs. I do not remember any P_{pk} requirements originating from customers. The Automotive and Aerospace industries were among the first to adopt C_{pk} requirements. These industries also noticed that in-spec products that also experienced OOC events during production often had early life reliability failures or other latent quality issues. (Note: As a Chief Quality Executive for multiple companies, where I also managed the Reliability Assurance and Customer Complaint Management Departments, I can personally confirm these observations.) So, they added a "Ship-to-Control" requirement in addition to the C_{pk} requirement of ≥ 1.67 for Critical-To-Quality (CTQ) Product Characteristics. Customers not only wanted their products to meet specifications but also wanted the process runs to be in control. This is certainly reasonable.

To address the "Ship-to-Control" imperative and mitigate the risk of unreliable products, three key industry standards were established: JEDEC JESD50C: Special Requirements for "Maverick" Product Eliminations and Outlier Management; and AEC-Q004: Automotive Zero Defects Framework; and SEMI C64: Statistical Guidelines for Ship to Control. (NOTE: SEMI C64 was last approved in 2014 and is now considered inactive.) Chapter 13 provides a detailed examination of effective strategies for meeting "Ship-to-Control" requirements through the application of these standards.

Suppliers often struggled to meet C_{pk} ≥ 1.67 requirements for all Product CTQs, leading to the creation of three categories: Critical, Key and Significant Characteristics as a compromise. Critical Product Characteristics were redefined as essential for safety and performance. Key Characteristics were defined as important but not safety related. Significant Characteristics were defined as other important CTQs for overall product quality. The C_{pk} goals for both Significant and Key Characteristics were often reduced to ≥ 1.33.

Even though customer-specified C_{pk} requirements could be addressed during contract negotiations, they led to additional burdens on suppliers. Most importantly, additional product screening requirements were imposed to make sure that only production units meeting specifications were shipped. These additional product screening requirements often led to place specifications on control charts which triggered competing priorities starting the *"domino effect."* Indeed, many customers asked that Specification Limits be added to control charts (recall discussion in 5.2). Even though customers had good intentions, as previously discussed in Chapter 4, the *"domino effect"* from these competing priorities often led to ineffective "SPC for Show" Programs.

10.2. Common C_{pk} Use Errors

In this section, I examine the most common errors when using Process Capability Index C_{pk} for Product CTQs.

REPORTING C_{pk} FOR INDIVIDUAL PROCESS RUNS

Frequently, organizations reported C_{pk} values for individual process runs for Product CTQs. This is often reinforced by SPC production software, where the software can easily be set up to report individual process run and/or production lot C_{pk}. However, recall the premise for a qualified control chart: a stable process continues to deliver the same average and variability unless there an OOC event is triggered by a special cause rule violation indicating a process shift. Remember you are taking a sample to determine if the process has shifted! So, the C_{pk} for each process run should be what was calculated for the whole control chart.

TAKING MORE SAMPLES TO ESTIMATE PROCESS RUN C_{pk}

Often, some organizations attempt to estimate process run C_{pk} by taking additional samples. The argument being more samples provide a better estimate of μ and σ. This approach works if the additional samples drawn are random and independent from each other. However, we know from *Effective Sample Size* (n_{eff}) principles, this is often not true. *Effective Sample Size* (n_{eff}) analysis ensures that the selected sampling plan for each process run represents the most random and independent sample selection. This ensures that the subgroup (process run) dataset most accurately reflects the current process variability. Taking more samples that are not random or independent will incorrectly calculate μ and σ, and by extension C_{pk}. This should be done only if it is confirmed that the additional samples are statistically random and independent.

In both cases, the only exception is if you measured the Product CTQ for all production units within the process run. For example, this is true in the semiconductor industry but only at Final Test. This industry's Final Test Operation measures all production units for defined Product CTQs. In this case, you have the complete dataset, μ , σ and by extension the process run C_{pk} will be accurate.

REPORT C_{pk} WHEN ITS REALLY P_{pk}

Recall that C_{pk} for Product CTQs assumes that the process is in statistical control, or as suggested in this book, has a *Stability Index* (S_i) ≥ .85. Many organizations calculate and report C_{pk} by combining the dataset from all equipment set up to deliver the same process, irrespective of their control chart behavior. This situation is often complicated by many production SPC software where default features automatically and incorrectly provide these calculations as C_{pk}. We now know that these calculations should be reported as P_{pk} if the data set has a *Stability Index* of < .85 as it represents the process performance for an unstable process. The good news is some production SPC software can be set up to clarify whether the calculation is C_{pk} or P_{pk}.

10.3. Summary of Recommendations

Improper use of process capability indices instead of using process performance indices can trigger the *"domino effect"* rendering ineffective control charts. My key recommendations include:

- Use the *Stability Index ≥ .85* to determine if you should report Capability Indices (C_p, C_{pk}) or Performance Indices (P_p, P_{pk}) .
- Do not invent product specifications for Process CTQs.
- Do not report C_{pk} for individual process runs unless you measure all units within the process run.
- Report P_{pk} when combining datasets from various equipment unless the *Stability Index* ≥.85.

Next, let's explore how to calculate the most reasonable product specifications.

11. CALCULATING REALISTIC PRODUCT SPECIFICATIONS

Design Engineering begins setting Product CTQs specifications during new product development. Many Design Engineering organizations collaborate with their Production counterparts to ensure product specifications are manufacturable during the design process. In other cases, such as tightening previously used specifications, they assume that Production will learn how to improve their processes to meet these new specifications. Some new products introduce features, technologies, or approaches that advance the state of the art in a particular field. Often, the manufacturability of these new specifications is still unknown.

New products are often developed and tested to evaluate designs without focusing on long-term manufacturability. Sometimes they are initially built in separate pilot lines before being transferred to the primary production line. Constructing new products according to the specifications is typically the initial focus. Successful designs were then transferred to production personnel for qualification in the primary production line. Product qualification methods focused on product performance and reliability using devices that meet specifications, as well as production readiness. Production readiness includes ensuring that products and processes are fully documented, specifications are established, operators are trained and certified, and process control activities are in place. Once qualified, the production team becomes responsible for improving the new product's manufacturability to meet process capability requirements (assuming they could not immediately be met) while design engineering proceeds to the next new product.

Production now focuses on improving and/or maintaining the manufacturability of new products. Statistical Process Control (SPC) methods are crucial for these improvements. Many companies voluntarily use SPC, while some do so only when required by customers or adopted quality standards. For Product CTQs with low C_{pk}, production works to improve performance through SPC, process adjustments, and/or experimentation. If unsuccessful, they implement additional screening methods, often lowering yields to ensure only in-spec products are shipped. This creates competing priorities and a "*domino effect,*" leading to ineffective SPC programs (see Chapter 4).

I propose two methods to improve this situation. First, use C_{pk} to determine the "best you can do" specifications when applying SPC methods to improve and stabilize all equivalent equipment. Second, use P_{pk} formulas to determine specifications that match your current shipments for all equipment delivering the same process.

11.1. Use C_{pk} to Calculate BEST - YOU - CAN - DO Product CTQ Specifications

This method uses C_{pk} to calculate candidate specifications. Here we use the same methods and assumptions described in Chapter 10. You can/should use data from *qualified* (as defined in Chapter 7.15), stable, *well-behaved* process control charts that pass the 85% Rule (*Stability Index* ≥ *.85*). You can include data from equivalent equipment when you combined them on a common control chart using the methods described in Chapter 8.

- Recall the C_{pk} Formula for Double-Sided Specification Limits:

$$C_{pk} = \min\left(\frac{USL-\mu}{3\sigma_{within}}, \frac{\mu-LSL}{3\sigma_{within}}\right)$$

- Let's make some edits to the equation and re-define the variables:

$$C_{pk} = \min\left(\frac{USL_C-\mu}{3\sigma_{within}}, \frac{\mu-LSL_C}{3\sigma_{within}}\right)$$

Where: μ = dataset average
$\sigma_{within} = \frac{\bar{s}}{c_4}$
$\bar{s}$ = average of subgroup standard deviations
c_4 = correction factor for subgroup sample size from Table 36
C_{pk} = desired or specified Process Capability
USL_c = Candidate Upper Specification Limit
LSL_c = Candidate Lower Specification Limit

- Now let's re-arrange the equation and solve for calculating the Upper Spec Limit:

$$USL_c = \mu + (C_{pk} \times 3\sigma_{within})$$

- Similarly, this is the equation for calculating the Lower Spec Limit:

$$LSL_c = \mu - (C_{pk} \times 3\sigma_{within})$$

Note: You can use the same equations for situations when you have a single sided spec.

Consider the control chart in Figure 23 where all plotted data points appear in control. This chart has a *Stability Index* = 1.00, so it passes the 85% Rule. So, we can use the dataset to calculate candidate specifications.

Figure 23 Qualified Chart for Calculating Specifications

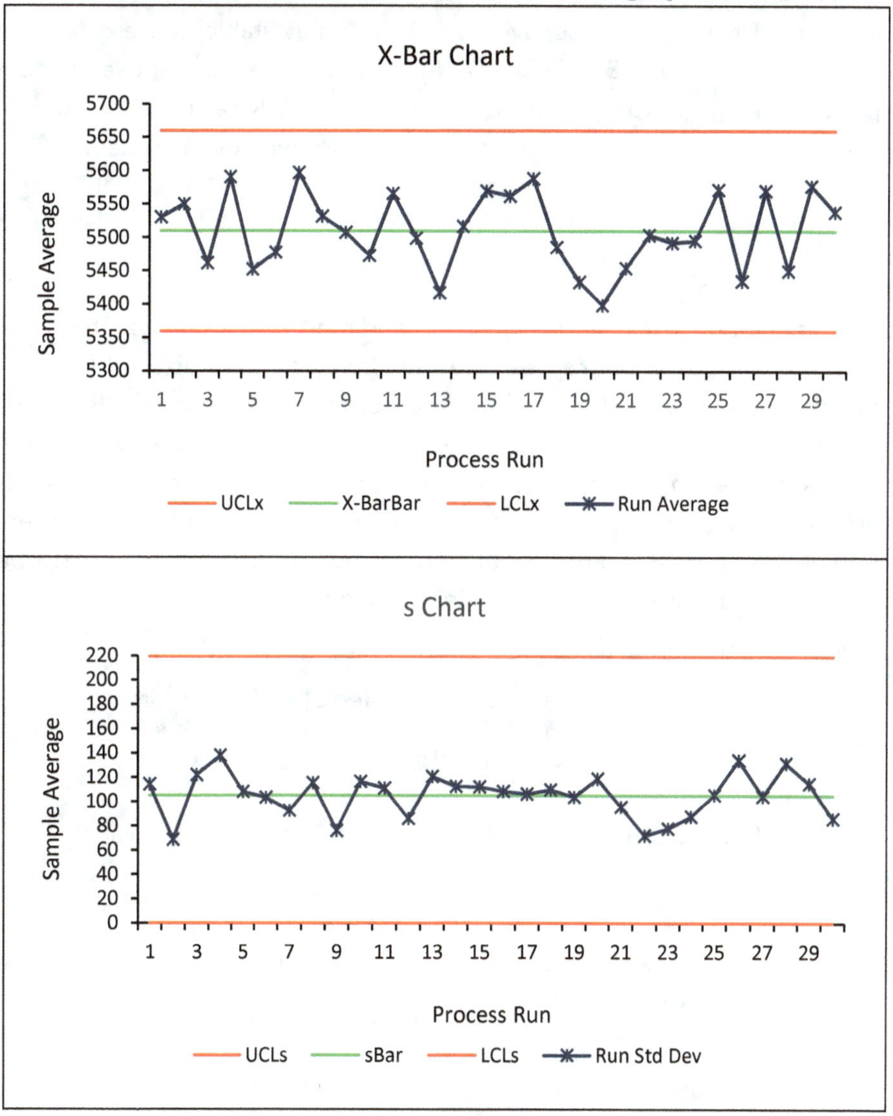

From the underlying dataset, we find:

$$\mu = 5510$$

$$\sigma_{within} = 107$$

Further, let's assume our organization and/or our customers require a $C_{pk} \geq 1.67$. Finishing our calculations:

$$USL_c = 5510 + ((3 \times 107) \times 1.67)) = 6046$$

$$LSL_c = 5510 - ((3 \times 107) \times 1.67)) = 4974$$

So how do I interpret these results? Given that you include all equivalent equipment that are well-behaved and reasonably stable, these candidate specifications are the BEST-YOU-CAN-DO with that equipment set without process changes. Now you can compare these newly calculated candidate specifications with current specifications to determine the next steps.

11.2. Use P_{pk} to Calculate BEST - MATCH - YOUR - SHIPMENTS Product CTQ Specifications

This method uses P_{pk} to calculate candidate specifications. The same methods and assumptions described in Chapter 10 are applied here. P_{pk} does not require a stable or *well-behaved* process, so passing the 85% Rule is not necessary for its calculation. P_{pk} data can include all process equipment delivering the same Product CTQ regardless of their qualification status. It is also acceptable to incorporate performance data from all equivalent equipment processes from various control charts. They only need to be configured to deliver the same process and Product CTQ.

Here are the formulas:

- Recall the P_{pk} Formula for Double Sided Specification Limits:

$$P_{pk} = \min\left(\frac{USL-\mu}{3\sigma_{overall}}, \frac{\mu-LSL}{3\sigma_{overall}}\right)$$

- Let's make some edits to the equation and re-define the variables:

$$P_{pk} = \min\left(\frac{USL_C-\mu}{3\sigma_{overall}}, \frac{\mu-LSL_C}{3\sigma_{overall}}\right)$$

Where: μ = dataset average
$\sigma_{overall}$ = standard deviation of all sample measurements
P_{pk} = desired or specified Process Performance
USL_c = Candidate Upper Specification Limit
LSL_c = Candidate Lower Specification Limit

- Now let's re-arrange the equation and solve for calculating the Upper Spec Limit:

$$USL_C = \mu + (P_{pk} \times 3\sigma_{overall})$$

- Similarly, this is the equation for calculating the Lower Spec Limit:

$$LSL_C = \mu - (P_{pk} \times 3\sigma_{overall})$$

Note: You can use the same equations for situations when you have a single sided spec.

Consider the control chart in Figure 24. This figure depicts a control chart with 12 out of 30 data points OOC failing Rule #1. This chart has a *Stability Index* = .40, so it obviously fails the 85% Rule. Indeed, it shows recurring process shifts. Yet we can still use the dataset to calculate candidate specifications using P_{pk}.

Figure 24 Example Stability Index = .40

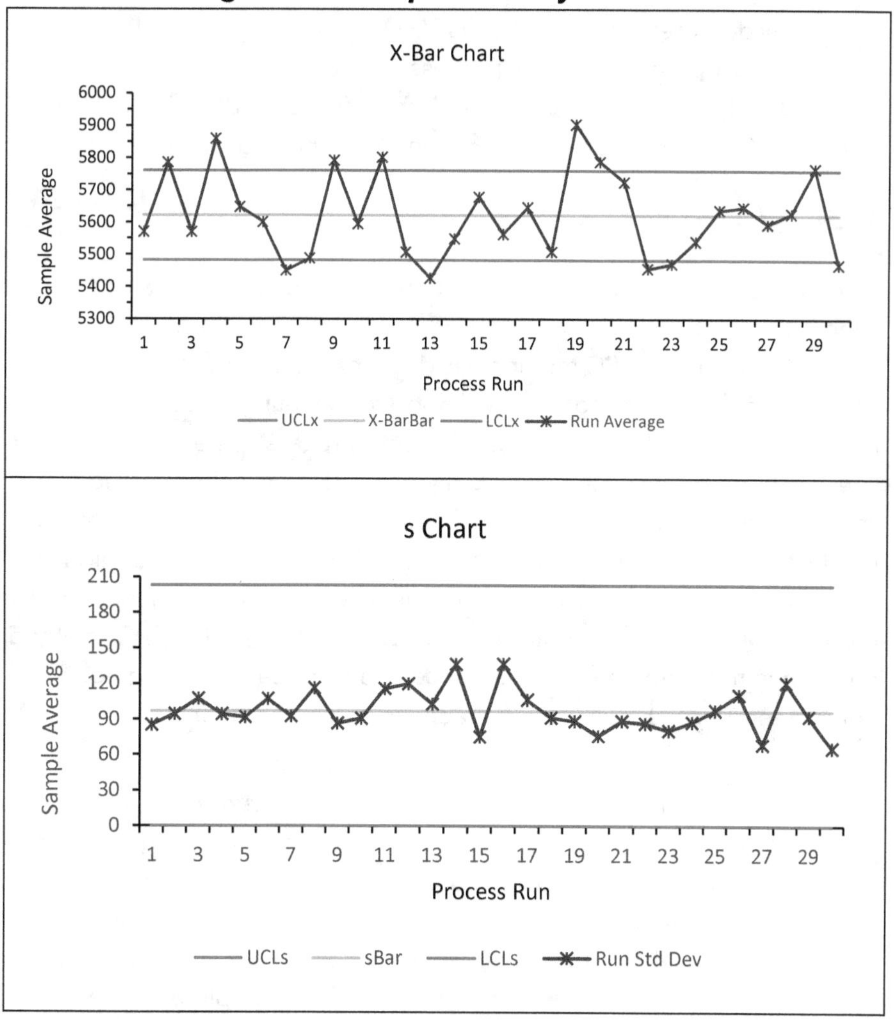

From the underlying dataset, we find:

$$\mu = 5623$$

$$\sigma = 161$$

Further, let's assume our organization and/or our customers require a $C_{pk} \geq$ 1.67. Finishing our calculations:

$$USL_c = 5623 + ((3 \times 161) \times 1.67)) = 6427$$

$$LSL_c = 5623 - ((3 \times 161) \times 1.67)) = 4818$$

So how do I interpret *these* results? Given that you include all equipment intending to deliver the same process, irrespective of whether they are stable, these calculated candidate specifications BEST-MATCH-YOUR-SHIPMENTS with your current equipment set without further improvements.

11.3. How to use Candidate Specifications

How can candidate specifications be utilized? C_{pk} candidate specifications demonstrate the potential performance if the process improvements using SPC methods as described in this book are successful. This allows for assessing the likelihood of meeting specifications if all equipment can deliver an equivalent process. The P_{pk} candidate specifications reflect your current delivery performance. By aligning current specifications with actual performance data, the costs associated with product screening or reworking can potentially be reduced. Additionally, it will lessen the need for process adjustments or other experimentation. The organization can concentrate their efforts on using the SPC methods outlined in this book to further enhance and stabilize all existing equivalent equipment. This focus will considerably decrease competing priorities.

There is another action you can take with the candidate specifications. *Approach Design Engineering to change the current Product CTQ specifications to one of the candidate specifications!* To my surprise, when I used this approach, *Design Engineering approved the candidate specifications ~ 80% of the time!* I am always amazed at that success rate. Often, they explained that their specifications were estimates, or that samples during development did not test the complete range of the specification limits because they were not available. They further explained that during product qualifications, they purposely chose samples for reliability testing close to the center of specification. When asked to consider the candidate specifications they often asked for product samples at the candidate specification limits and tested them both for performance and reliability before approval.

When this happened, production's focus changed. It became more important to stabilize the process which significantly reduced competing priorities described in Chapter 4. Processes stabilized quicker, process yields significantly improved and, of course, C_{pk} improved. I do not remember any downsides!

12. AUDITING SPC PROGRAMS FOR EFFECTIVENESS

As outlined in Chapter 3, the implementation of a Statistical Process Control (SPC) program can provide significant benefits, especially within manufacturing, service industries, and other quality-focused environments. These programs are generally adopted and implemented for two main reasons. Many SPC Programs are implemented voluntarily based on the anticipated improvements they can provide. Some SPC Program are implemented to comply with requirements such as those specified by an adopted Quality Standard or as mandated by customers. For example, certain quality standards necessitate SPC programs such as the automotive IATF 16949 standard. SPC is recognized by the IATF 16949 standard as one of the "Core Tools" in an automotive quality management system, alongside APQP, PPAP, FMEA, and MSA.

Customers or 3rd Party Auditors frequently audited my Statistical Process Control (SPC) programs to ensure they met their requirements. These audits were conducted by their Supplier Quality Engineers or outsourced to 3rd Party Audit Agencies. I observed that many of these external evaluations, although not all, offered limited value as they primarily focused on confirming the existence of SPC Programs rather than assessing their effectiveness. This was particularly true for compliance audits to Quality Standards. Fortunately, when this was their focus, findings for my organization were infrequent.

However, some customer audits identified findings based on their specific requirements, some of which undermine effective SPC Programs. For instance, requiring specification limits for individuals be placed on control charts for averages or mandating process adjustments for Out-Of-Specification (OOS) samples, even if the process adhered to special cause rules (i.e., the process was in control). Frequently, corrective actions addressing these findings lead to the "Competing Priority" conflict described in Chapter 4. These corrective actions often shift the emphasis from "SPC for Improvement" to compliance, resulting in the emergence of "SPC for Show" programs.

Within the organizations I worked for, the Quality Department was responsible for conducting Statistical Process Control (SPC) Program evaluations. This responsibility included defining the evaluation methodologies. Once defined, these evaluations were performed by me, my Quality Engineers, my Internal Quality Auditors, and/or my Supplier Quality Engineers during supplier assessments. I evaluated both internal SPC Programs for the companies I worked for, as well those of our suppliers.

Assessing the effectiveness of SPC Programs is essential to ensure they achieve their intended objectives and add value to the company. The assessment of SPC Program effectiveness requires more than merely confirming their existence. It should identify key weaknesses and provide recommendations for improvement. Well designed and comprehensive assessments provide improvement guidance to the Control Chart Owner (typically the primary Process Engineer) and ultimately the organization. When the Control Chart Owner and the organization took actions based on my assessments, it ensured that the SPC Program achieved its goals, including improving process stability, capability, reducing costs, and maintaining compliance with adopted Quality Standards.

The assessment methodologies outlined in this chapter consolidate my knowledge and insights that I acquired throughout my career. These methods offer a comprehensive yet straightforward approach for evaluating SPC Programs. They incorporate all the principles detailed in this book. These methodologies can be utilized to assess individual SPC Charts as well as complete SPC Programs within your organization as well as those at your suppliers. They are significantly more effective than traditional SPC evaluations that focus primarily on the existence of SPC. Additionally, they specifically diagnose situations that undermine the effectiveness of the SPC Program and provide the organization with an improvement roadmap.

These methods involve two essential steps. First, individual control charts are selected and scored. This step offers insights into the effectiveness of each control chart and provides diagnostic information for each one. Second, the scores from each chart are combined to produce a final score for the organization-wide SPC program. This step evaluates the overall effectiveness of the SPC program and allows for comparative scoring across multiple SPC programs in various organizations.

12.1. Assessing the Effectiveness of Individual Control Charts

The method begins by identifying and selecting at least 20 different control charts drawn from various process owners across the various production processes. These should include both Process CTQs and Product CTQs, with each control chart having a minimum of the *most recent* 50 process runs plotted on it. Selecting at least 20 different control charts ensures sampling a wide range of process owners. Be sure to select Control Charts across manufacturing, sample different control chart owners, and different process engineers as well as different engineering departments. This helps diagnose whether an organization has a chronic problem or if ineffective control charts are linked to specific control chart owners and/or departments. Ensuring control charts plot a minimum of 50 process runs makes it easier to determine if the control chart is *well-behaved* (see 7.12), the control limits are reasonable, whether the chart is plotting process runs versus production lots, or whether OCAPs are being effectively applied.

Once the candidate charts are identified, the scorecard criteria described in Table 27 (repeated as Appendix 15.5) will be applied to each individual control chart. (This table can easily be reconstructed as an Excel workbook.) For each finding, points will be added or subtracted as specified in the table. The point allocation adjustment is based on my experience as to how much the criteria affects the effectiveness of the SPC control chart. The criteria implement all the key learnings described in this book.

Each control chart Scorecard begins at 100 points, with adjustments made according to the criteria table. Positive adjustments reflect observed best practices, while negative ones indicate observed ineffective practices. Higher scores mean more effective control charts; lower scores mean less effective control charts. Final scores range from 0 (worst practices) to 200 (best practices). Each Control Chart Scorecard evaluates chart effectiveness, so sharing this information with the Control Chart Owner or Process Engineer helps them improve its performance.

Table 27 Individual Control Chart Scorecard

Auditor:			**Date:**	/ /
CTQ Name:			**Product CTQ:**	☐
Chart Owner:			**Process CTQ:**	☐
Process Engineer:			**Starting Score**	**100**
	Criteria	**Possible Adjustment**	**Actual Adjustment**	**Score**
1	Control Limits Appear Unreasonable ☐ Appear too tight ☐ Appear too wide ☐ Other____________	-10		
2	Specification Limits Plotted on Control Chart	-10		
3	Specification Limits are defined for a Process CTQ	-10		
4	Control Chart plots Product Lots versus Process Runs (exception, no deduction if 1 production lot = 1 process run)	-10		
5	OCAPs are not clearly documented	-10		
6	OCAPs specify investigation of the OOC Root Cause	+20		
7	OCAPs specify consistent adjustment	+10		
8	OCAPs allow inconsistent adjustments	-10		
9	OCAP calls for adjustment when OOS sample detected	-10		
10	OSAPs & OCAPs clearly different	+25		
11	OSAPs for Product CTQs are not clearly documented	-10		
12	OSAP calls for adjustment when OOS sample detected	-10		
13	OSAPs calls for containment, product disposition or rework the OOS material, no process adjustments are allowed	+20		
14	Able to Verify OCAP execution for all OOC events	+25		
15	Cannot verify OCAP execution for all OOC events	-10		
		Control Chart Final Score:		
Key Definitions				

OCAP = Out of Control Action Plan OSAP = Out of Specification Action Plan CTQ = Critical To Quality	OOC = Out Of Control OOS = Out Of Specification

Let's review each of the criteria!

12.1.1. Control Limits Appear Unreasonable

This criterion can be applied by quickly reviewing the plotted control charts and is designed to assess whether the control limits appear reasonable for the data plotted. Most of the time, you only need to apply this criterion to the X-Bar chart (assuming an X-Bar, s-Chart).

First, let's examine what a reasonable control chart (Figure 25) should look like.

Figure 25 Chart with Reasonable Control Limits

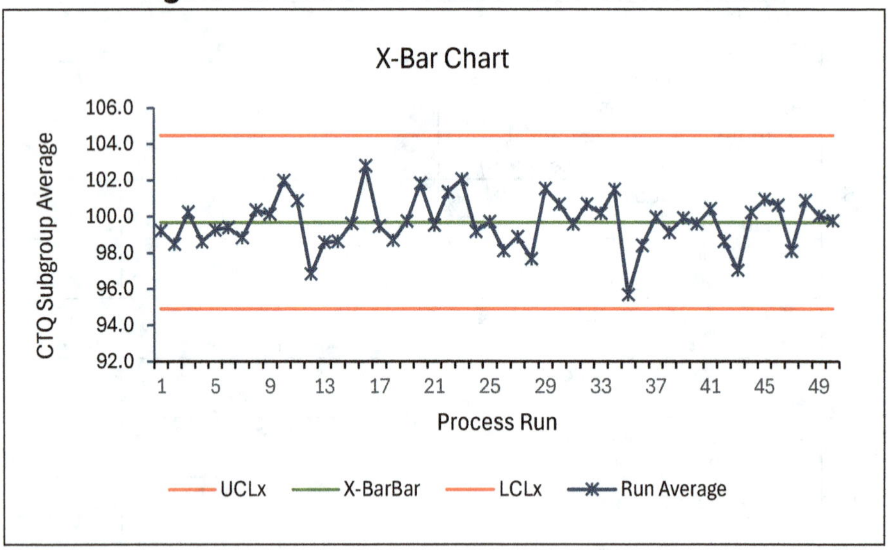

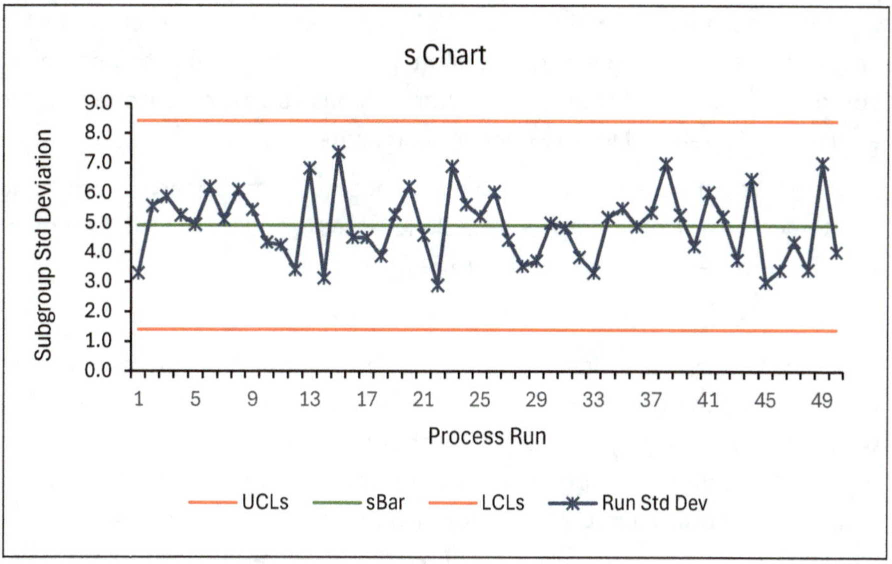

The most important feature to look for at this point is the distribution of plotted run data. They should completely fill the control limits on both the X-Bar and the s-Chart. Indeed, if you were to create a histogram for the X-Bar data (Figure 26), according to the Central Limit Theorem it should look like a normal distribution with the data point tails close to the upper and lower control limits.

Figure 26 Distribution of Process Run Averages

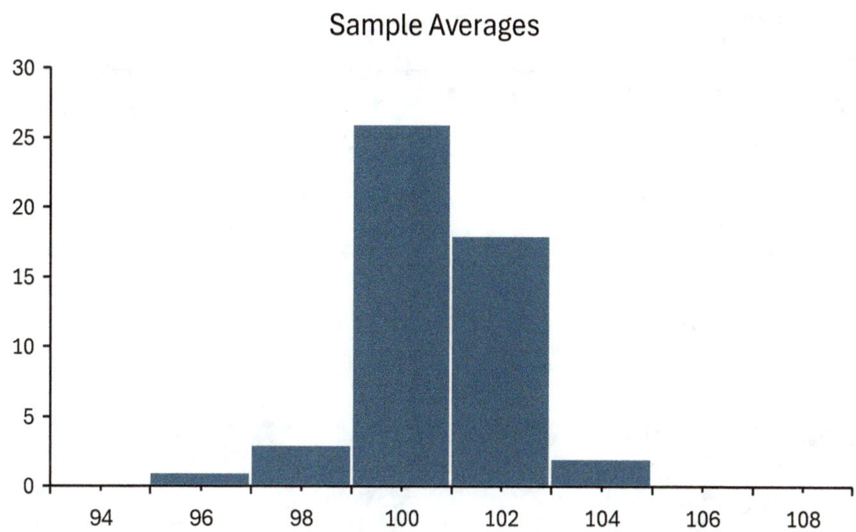

In this example, the X-Bar UCL is 104.4 and the X-Bar LCL is around 94.9. The histogram indicates a distribution tail around the upper and lower control limits. Conclusion: the control limits appear reasonable. No adjustment is needed for the control chart scorecard.

Now let's consider the two most frequent cases that I observed indicating that the control limits are unreasonable for the plotted data. I label them as: *Too Good to be True*; and *Too Bad to Be True.*

Too Good to be True

Figure 27 depicts a classic example of Too Good to Be True control limits for the X-Bar Chart. I observed this type of chart at multiple suppliers. They were often very happy because they thought it implied a very controlled process. In reality, the plotted X-Bar dataset is just too tight for the control limits; or the control limits are just too wide for the plotted dataset.

Figure 27 Too Good to be True

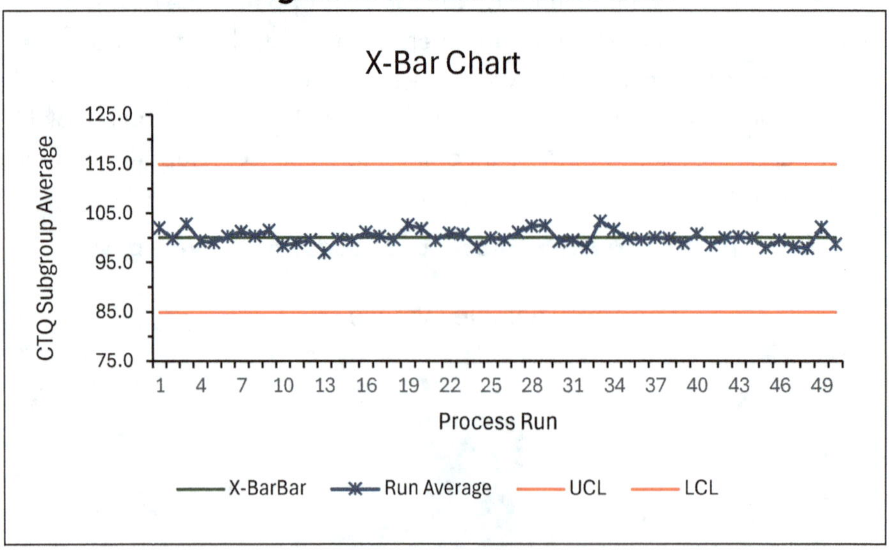

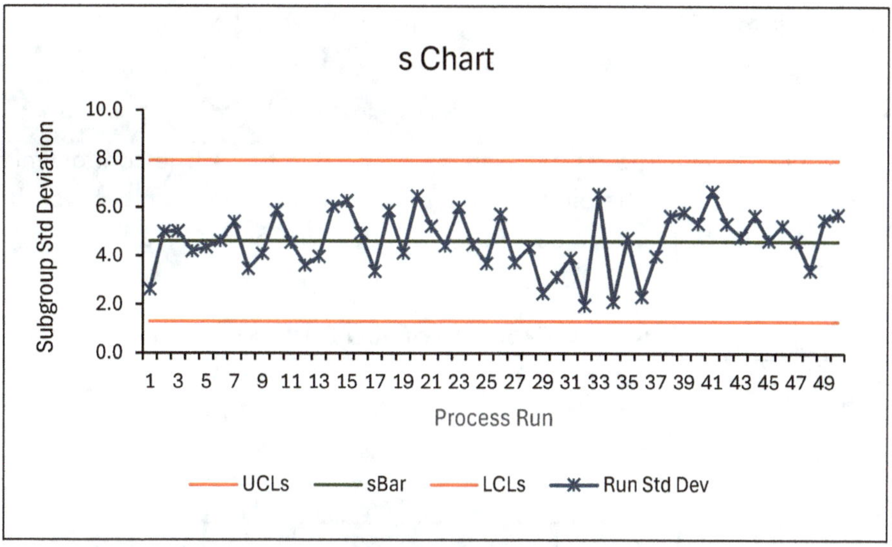

Recall, that if you were to create a histogram for the X-Bar data, according to the Central Limit Theorem it should look like a normal distribution (Figure 28) with the data point tails close to the upper and lower control limits.

Figure 28 Distribution of Sample Averages

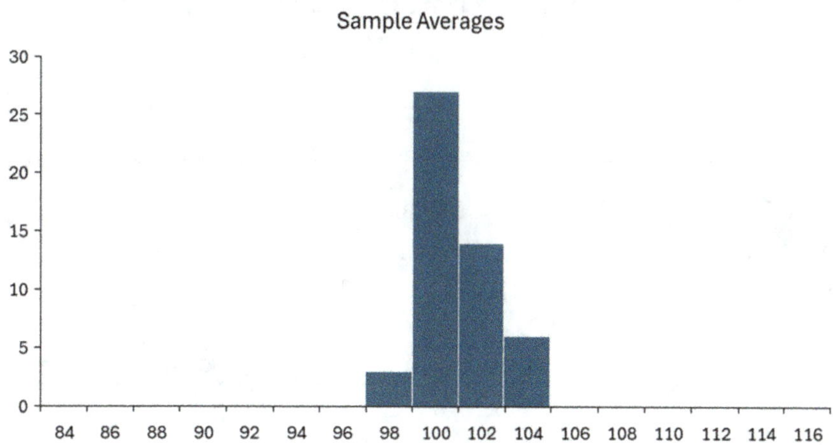

Notice the dataset tails are nowhere near the control limits of 85 and 115. Conclusion: the control limits appear too wide. In this case you would deduct 10 points (-10) from the control chart scorecard.

My experience suggests that the number one reason for this situation was that the control limits were actually specification limits and not calculated control limits. Unbelievably, the number 2 reason (primarily found at suppliers) for this situation was the control limits were artificially set in order to imply a well-controlled process.

Too Bad to be True

Figure 29 depicts a classic example of Too Bad to Be True control limits for the X-Bar Chart. The plotted X-Bar control limits appear too tight for the dataset. Indeed, it fails the 85% Rule introduced in Section 7.12.

Figure 29 Example of Too Bad to be True

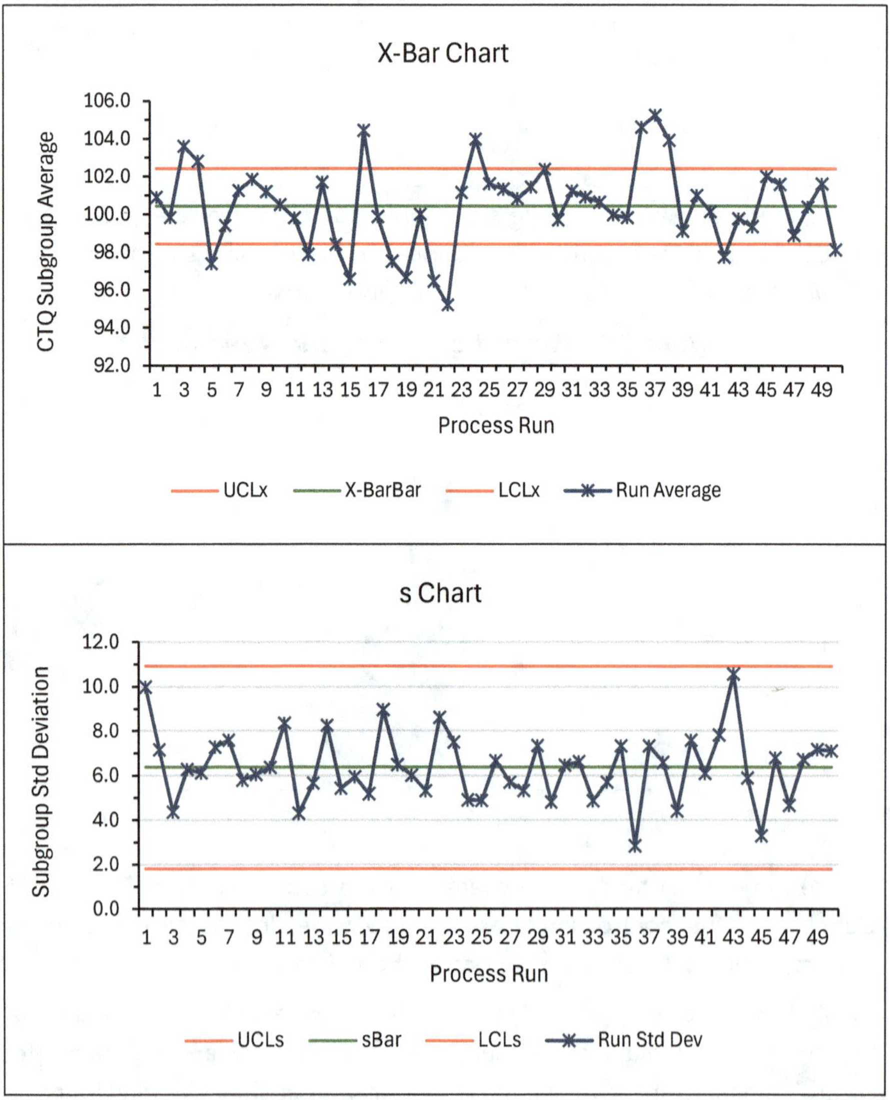

The histogram of process run averages is depicted in Figure 30. Notice the distribution of averages fall well past the control limits of 102.4 and 98.6.

Figure 30 Distribution of Process Run Averages

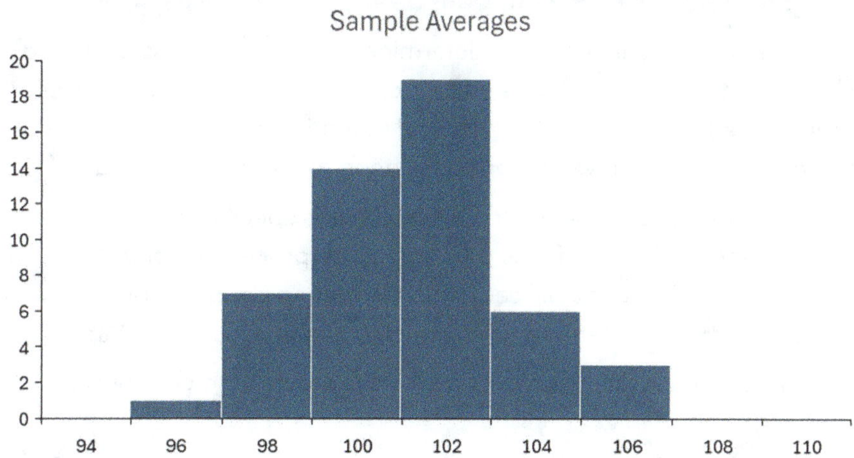

Conclusion: the control limits appear too tight. In this case you would deduct 10 points (-10) from the control chart scorecard.

Experience shows the most frequent root causes for this situation are the control chart is setup with the wrong sample size (see Section 5.3), or the process is being affected by *Lurking Variable* (see Section 5.4), or, "Copy Similar" equipment were combined on the control chart too early (see Section 8.2).

The impact of this situation was that OCAPs, or adjustments were made more often than needed, which negatively impacted both process capability and process stability.

12.1.2. Specification Limits Plotted on the Control Chart

It is easy to determine whether specification limits are on a control chart as they are usually labelled as such. Just ask the Chart Owner or the Process Engineer. When you observe this situation deduct 10 points (-10) from the control chart scorecard. I only deduct 10 points since I have additional criteria used to determine if having these Specification Limits on the control chart is hindering the possible benefits of the SPC Chart. If they hinder the effectiveness of the SPC Chart, more points will be deducted.

Recall, Control Charts and their Control Limits are for averages (exception: an individual/moving range chart) while specifications are for individual units. So why would you place limits for individuals on a chart for averages? See Section 5.2.

12.1.3. Specification Limits are defined for a Process CTQ

This situation is also easy to determine. Have the process owner advise whether the monitored CTQ is a Product CTQ or a Process CTQ (check the appropriate box in the Scorecard) as defined in Table 1 in this book, then determine whether specifications have been defined for this CTQ.

Recall from Chapter 4, do not invent process specifications when you do not need them. Maintain focus on keeping the process in control. Inventing unneeded specifications will cause the competing priority of product versus process control and start the *"domino effect"* as described in Chapter 4.

If this situation exists, deduct 10 points (-10) from the control chart score card.

12.1.4. Control Chart plots Product Lots versus Process Runs

This situation is usually easy to determine by asking the Chart Owner or Process Engineer. Often it is as simple as reading the x-axis label on the control chart. If you find the x-axis is a production lot versus a process run, deduct 10 points (-10) from the control chart scorecard. The only exception is if one process run is also a one production lot. See Section 5.1.

12.1.5. OCAPs are not clearly documented

Investigating this criterion is a little more difficult because it cannot be determined by looking at a control chart. This requires locating the governing document for the specific CTQ being monitored and determining if the documentation *clearly* specifies what Out-of-Control Actions Plan (OCAPs) should be taken should an Out-of-Control (OOC) event be triggered. Usually, it is found within the approved work instructions governing the process. Sometimes it is a separately approved document. Sometimes it is only documented in the SPC production software. Be on the watch for global SPC documentation that affects all SPC charts as these often suggest generalized OCAPs that are not Process CTQ specific.

If specific Process CTQ OCAP documentation does not exist, or they are general OCAPs across multiple CTQs, subtract 10 points (-10) from the SPC scorecard.

12.1.6. OCAPs specify investigation of the OOC Root Cause

If the OCAP documentation includes provisions for investigating the root cause of the OOC event, this is considered the best practice. Therefore, you add 20 points (+20) to the scorecard. When this approach is effectively used it usually results in the most stable processes and improves process capability.

12.1.7. OCAPs specify consistent adjustments

If the documentation specifies consistent process adjustments as the OCAP meaning all personnel across all shifts perform the same adjustment when an OOC occurs, add 10 points (+10) to the scorecard. While it is not as effective as a root cause investigation OCAP, performing consistent adjustments quickly determines the OCAP effectiveness and ultimately leads to process improvements.

12.1.8. OCAPs allow inconsistent adjustments

Deduct 10 points (-10) from the scorecard if OCAP documentation permits inconsistent adjustments or leaves decisions to individuals during OOC events. For example, letting process operators choose adjustments, or only instructing them to contact the on-duty Process Engineer for decisions. The on-duty Process Engineer is then authorized to decide on what adjustment to implement. My experience suggests inconsistent process adjustments often degrade process performance. Further, it makes it very difficult to determine the actual root cause for the OOC event. This approach is often the root cause for the "*domino effect*" (see Chapter 3).

12.1.9. OCAP calls for adjustment when OOS Sample detected

When an OCAP calls for or allows a process adjustment whenever an Out-of-Specification (OOS) (independent of an OOC event) sample is detected is a classic case of setting up "Competing Priorities" as described in Chapter 4. Of course, if these actions are documented, it is easy to find and potentially change.

Most of the time in this situation, process stability and process capability degrade.

When you diagnose this situation subtract 10 points (-10) from the scorecard.

12.1.10. OSAPs & OCAPs are clearly different

This is clearly a best practice, so when you confirm Out-of-Specification Action Plans (OSAPs) and Out-of-Control Actions Plans (OCAPs) *are clearly documented and are clearly different* award 25 points (+25) to the control chart scorecard. This is the first step in preventing competing priorities described in Chapter 4.

12.1.11. OSAPs for Product CTQs are not clearly documented

When you cannot find clear documentation for OSAPs for Product CTQs, deduct 10 points (-10) from the control chart scorecard. My experience revealed that in this situation either process adjustments occur or the affected out of specification product is ignored, i.e., the affected product is not contained.

12.1.12. OSAP calls for adjustment when OOS

If you find that the OSAP documentation calls for an adjustment when an OOS event is detected, you just found a key root cause for competing priorities causing ineffective control charts (Chapter 4). The results of these actions are very similar to *OCAPs calls for adjustment when OOS sample is detected:* Competing priorities emerge rendering ineffective control charts.

Deduct 10 points (-10) from the control chart scorecard.

12.1.13. OSAPs calls for containment, product disposition or rework the OOS material, no process adjustments are allowed

This is absolutely the best practice when approved documentation includes prescribed actions for the containment, product disposition, or rework of detected OOS material. In addition, if the documented action plan does not allow process adjustment solely based on OOS event, it means that the control chart owner has eliminated competing priorities. Award the control chart scorecard 20 Points (+20) but again only if process adjustments are not allowed when OOS samples are detected.

12.1.14. Able to Verify OCAP execution for all OOC events

This is often the most difficult audit action as it requires that you: verify that the Control Chart detected an OOC event, confirm that OCAPs are fully documented in *approved documentation*; and confirm actual actions taken match the approved documentation.

First verify which SPC Rules were enabled for the control chart (this is often enabled in the SPC Software). Have the Control Chart Owner identify the triggered OOC events for the control chart being audited. Specifically look for Rule 1: One point out-of-control rule, or Rule 4: 8 consecutive points on either side of the centerline (Section 7.15) violations. Recall, these are the most evident OOC Rules and the easiest for an auditor to detect when reviewing a control chart. They are also the most effective OOC Rules.

For OOC triggered events, then investigate if actual actions taken were fully documented. Finally, match actions taken against expected actions found in approved documentation.

When they all match, it is a Best practice and award the control chart scorecard 25 points (+25).

12.1.15. Cannot verify OCAP execution for all OOC events

However, if you cannot verify the execution of an OCAP when an OOC event occurs subtract 10 points (-10) from the control chart scorecard. Deduct 10 points even if actions were taken but they were different than the documented and approved OCAP actions; or there is no documentation of actions taken; or some OOC events did not trigger OCAP actions.

12.2. Interpreting an Individual Control Chart Score

Once you complete your audit and you have the final score, you should share your results with the Control Chart Owner and discuss the results. Table 28 provides a high-level interpretation of control chart assessment scores.

Table 28 Control Chart Score Interpretation

Score	Interpretation	Control Chart Effectiveness
200	• Highest Possible Score. • The Control Chart follows the best practices described in this book. • There are no identified factors that are reducing its effectiveness. • The Control Chart is *well-behaved.*	• The Control Chart is considered highly effective. • Expect process stability and capability to improve over time.

Score	Interpretation	Control Chart Effectiveness
>100	• Some best practices described in this book have been implemented.	• The Control Chart is considered effective. • Expect process stability and capability to improve over time.
>60	• Some identified factors are limiting the Control Chart's improvement potential.	• The Control Chart is somewhat effective. • Expect process stability to remain the same or degrade slowly over time. • Expect process capability to remain the same or degrade slowly over time. • Do not expect improvement.
<60	• Key factors are limiting the effectiveness of the control chart.	• The Control Chart is most likely not effective. • Expect process stability and capability to degrade over time.
0	• Lowest Possible Score. • Critical factors exist that are limiting the effectiveness of the control chart.	• The Control Chart is not effective. • Expect process capability and stability to quickly degrade over time.

Even better, the scorecard provides direct feedback to the Control Chart Owner on how to improve the effectiveness of their control chart. Table 29 provides an example of a completed control chart score card.

Table 29 Example of a Completed Scorecard

Auditor:		Steven Bergeron	**Date:**		10/04/2026
CTQ Name:		Die Shear	**Product CTQ:**		☒
Chart Owner:		Joe Engineer	**Process CTQ:**		☐
Process Engineer:		Joe Engineer	**Starting Score**		**100**
	Criteria		**Possible Adjustment**	**Actual Adjustment**	**Score**
1	Control Limits Appear Unreasonable ☐ Appear too tight ☐ Appear too wide ☐ Other____________________ _____		-10	0	100

2	Specification Limits Plotted on Control Chart	-10	-10	90
3	Specification Limits are defined for a Process CTQ	-10	0	90
4	Control Chart plots Product Lots versus Process Runs (exception, no deduction if 1 production lot = 1 process run)	-10	-10	80
5	OCAPs are not clearly documented	-10	0	80
6	OCAPs specify investigation of the OOC Root Cause	+20	0	80
7	OCAPs specify consistent adjustment	+10	0	80
8	OCAPs allow inconsistent adjustments	-10	-10	70
9	OCAP calls for adjustment when OOS sample detected	-10	-10	60
10	OSAPs & OCAPs clearly different	+25	0	60
11	OSAPs for Product CTQs are not clearly documented	-10	-10	50
12	OSAP calls for adjustment when OOS sample detected	-10	0	50
13	OSAPs calls for containment, product disposition or rework the OOS material, no process adjustments are allowed	+20	0	50
14	Able to Verify OCAP execution for all OOC events	+25	0	50
15	Cannot verify OCAP execution for all OOC events	-10	-10	40
			Control Chart Final Score:	40

Key Definitions	
OCAP = Out of Control Action Plan OSAP = Out of Specification Action Plan CTQ = Critical To Quality	OOC = Out Of Control OOS = Out Of Specification

Review Table 30 for the scorecard's findings and recommended improvement actions based on Table 29 Example of a Completed Scorecard.

Table 30 Example - Interpreting Score Card Findings

Criteria	Findings	Recommended Actions
2	• Specification Limits Plotted on Control Chart.	• Remove Spec Limits from Control Chart.
4	• Control Chart plots Product Lots versus Process Runs.	• Change the Control Chart to plot process runs. • Recalculate Control Limits.
8	• OCAPs allow inconsistent adjustments. • Specifically, the OCAP says to stop the process, call the on-duty Process Engineer, who can determine what adjustment to make.	• Specify consistent adjustments for all on-duty Process Engineer on all shifts. • Add provisions for OOC root cause analysis to the OCAP.
9	• OCAP calls for adjustment when OOS sample detected.	• Remove all OCAP adjustments related to OOS event. • Add OSAP instructions requiring product disposition.
11	• OSAPs for Product CTQs are not clearly documented.	• Document OSAPs that require product disposition, containment, or rework for OOS material. • Make sure OSAPs do not include process adjustments.
15	• Cannot verify OCAP execution for all OOC events.	• Ensure that all actions taken for an OOC event are fully documented and readily retrievable.

12.3. Assessing the Overall SPC Program's Effectiveness

To assess an organization's SPC Program, audit many Control Charts, CTQs, chart owners, and department owners. My experience delivering SPC training showed that adoption varied; those using my methods achieved effective control charts and enjoyed significant benefits. Many engineers cited *competing priorities* as a key reason for not adopting these methods, preferring OOS reactive adjustments over OOC proactive investigations. Sometimes process engineers were willing to implement my methods but were blocked by management (again for *competing priority* reasons). Nobody challenged my methods; it was all about *competing priorities*. Regular audits allowed me to identify non-adopters and demonstrate performance differences, helping influence others to adopt these practices.

So, my recommendation for an initial SPC Program Audit is to audit at least 20 distinct control charts characterized by 20 different CTQs, different control charts owners, and different department owners. You are certainly welcome to audit more. Table 31 provides an example of a SPC *Program's* Scorecard (repeated in Appendix 15.6). The table assumes a minimum audit of 20 control charts representing 20 different CTQs. Ensure that both the CTQ name and the corresponding control chart owner and department owner are documented. The selection of control charts and owners is critical–sampling should span across the facility, various chart owners, and multiple process engineering departments.

This information will later facilitate the identification of improvement opportunities. Record each Control Chart Score and calculate the average Control Chart Score as part of your analysis.

Table 31 SPC Program Scorecard

Control Chart	CTQ Name	Control Chart Owner	Department Owner	Score
1				
2				
3				
4				
5				
6				
7				
8				
9				
10				
11				
12				
13				
14				
15				
16				
17				
18				
19				
20				
			Average Control Chart Score:	

Table 32 provides guidance on how to interpret the results from this SPC Audit with respect to the organization's overall SPC Program Effectiveness. Column 1 contains the Average Control Chart score. Column 2 provides a general interpretation of the average; and Column 3 indicates whether the SPC Program being reviewed passed the SPC Effectiveness Audit using the principles and criteria described in this book.

Table 32 Interpretation of SPC Scorecard Average

Average Control Chart Score	**Interpretation of SPC Program Effectiveness**	**Audit Results**
>100	• The SPC Program is judged to be very effective. • Most Control Charts audited follow the best practices described in this book and are considered *well-behaved.* • You can expect to see improved process stability and capability.	Pass
≤80 but <100	• The SPC Program is judged to be reasonably effective. • Some control charts are *well-behaved.* • You can expect to see improved process stability and capability on high scoring control charts. • Locate the low scoring Control Charts and address the identified findings.	Pass • Issue minor finding on any Control Chart with score <70.

Average Control Chart Score	**Interpretation of SPC Program Effectiveness**	**Audit Results**
≥60 by < 80	• The SPC Program is judged to be partially effective. • Some control charts are *well-behaved.* • Some Control Charts are likely reasonably effective. • You can expect to see improved process stability and capability only on high scoring control charts. • Locate the low scoring Control Charts and address the identified findings.	Pass • Issue minor findings for each low scoring Control Chart.
<60	• The SPC Program is judged to be ineffective overall. • You should expect to see degrading process stability and capability.	Fail • Major finding issued for ineffective SPC Program. • Minor findings issued for each low scoring Control Chart.

Table 33 provides an example of an SPC Program Audit result using the techniques described in this book. First the average score for all the control charts audited is 80. The SPC Program is judged to be reasonably effective and therefore passes the Audit. You should also notice the patterns: Process_Engineer_3, 7 and 8 from Department C consistently had high scores, while Process_Engineer_2 and 5 both from Department B consistently had low scores.

Table 33 Example of a SPC Program Evaluation

Control Chart	CTQ Name	Control Chart Owner	Department Owner	Score
1	CTQ_1	Process Eng_1	Dept A	70
2	CTQ_2	Process Eng_1	Dept A	70
3	CTQ_3	Process Eng_2	Dept B	50
4	CTQ_4	Process Eng_2	Dept B	40
5	CTQ_5	Process Eng_3	Dept C	100
6	CTQ_6	Process Eng_3	Dept C	110
7	CTQ_7	Process Eng_3	Dept C	100
8	CTQ_8	Process Eng_4	Dept A	80
9	CTQ_9	Process Eng_4	Dept A	80
10	CTQ_10	Process Eng_4	Dept A	70
11	CTQ_11	Process Eng_5	Dept B	60
12	CTQ_12	Process Eng_5	Dept B	60
13	CTQ_13	Process Eng_6	Dept A	80
14	CTQ_14	Process Eng_7	Dept C	120
15	CTQ_15	Process Eng_8	Dept C	100
16	CTQ_16	Process Eng_8	Dept A	90
17	CTQ_17	Process Eng_8	Dept A	80
18	CTQ_18	Process Eng_8	Dept C	90
19	CTQ_19	Process Eng_9	Dept A	70
20	CTQ_20	Process Eng_10	Dept A	80
Average Control Chart Score:				80

Minor findings should be issued for each control chart with a score ≤70. The best practice is to provide each Process Owner with their individual scorecard.

SPC EFFECTIVENESS AUDIT RECAP

The assessment/audit described in this chapter provides organizations with feedback to help their SPC Program improve process stability, capability, cost reduction, and ongoing compliance with quality standards when findings are addressed.

When conducting an SPC Program Effectiveness Audit be sure to:

- Select 20 or more Controls Charts across manufacturing processes (be sure to sample the complete factory), include the most recent 50 process runs whenever possible. Be sure to select multiple Process and Department Owners.
- Assess each Control Chart independently and document findings using the criteria listed in Table 27 Individual Control Chart Scorecard.
- Use Table 31 SPC Program Scorecard to document the results from all the audited control charts.
- Use the criteria described in Table 32 Interpretation of SPC Scorecard Average to decide whether the SPC Program passes the effectiveness audit.

Provide feedback to the organization:

- Advise the organization on the Pass/Fail result of your audit.
- Provide each Control Chart Owner with a copy of their chart(s) Scorecard and explain each finding. Issue minor findings for each scorecard deduction by CTQ.
- Provide a copy of the overall SPC Program Assessment to the organization.
- Look for patterns such as by Control Chart Owner or Process Department. If patterns are found, advise the organization.

Be sure to follow through on their corrective actions to each finding to be sure they are suitably addressed.

13. THE MOST EFFECTIVE MAVERICK LOT PROGRAM

This chapter discusses three key situations, based on my experience, where I found a *specific* Maverick Lot Program to be particularly effective. First, addressing "Ship-to-Control" requirements. These could be customer-imposed requirements and/or requirements that you impose upon your suppliers. The second situation is when you are first implementing or upgrading an SPC program using techniques described in this book, and you want to "protect" your company during your SPC improvement journey. Finally, it is also very effective in situations when your company or one of your suppliers fails an SPC audit using the techniques described in Chapter 12; and you want to "protect" your company while your factory or your supplier address Audit findings.

13.1. Maverick Lot Program Overview

Let's begin by defining a "Maverick." A **Maverick Lot** is a production lot in which one or more key Product CTQ parameters show statistically significant deviation from historical or expected distributions, even though all measured values remain within datasheet or functional specifications limits. A **Maverick Unit** is an individual product that meets datasheet and functional specifications but exhibits statistically significant outlier behavior in one or more key parameters compared to the typical population.

The Maverick classification is important because, although these products may initially operate correctly, they carry a higher risk of early failure or reduced reliability, often signaling latent process excursions not captured by standard pass/fail criteria. My experience managing both product reliability and customer complaints supports this observation. As noted earlier, these findings have led customers to implement "Ship-to-Control" requirements in addition to C_{pk} requirements.

As suggested in Chapter 10, two active and complementary standards emerged to eliminate these unreliable products:

- JEDEC JESD50C: Special Requirements for "Maverick" Product Eliminations and Outlier Management. The JEDEC JESD50C standard defines how suppliers should identify, control, and disposition atypical lots/units to ensure high reliability products.

- The AEC-Q004 standard, developed by the Automotive Electronics Council (AEC) is a guide for semiconductor and electronic component manufacturers in achieving *zero-defect quality* for automotive applications, with stricter requirements tied to safety-critical applications.

They both emphasize proactive defect prevention, robust outlier detection, and continuous improvement across the supply chain. They are applicable in many industries including automotive electronics, semiconductor manufacturing, aerospace and defense, medical devices, and healthcare electronics as well as industrial and energy systems.

However, there are some key implementation strategies that can render Maverick Programs ineffective and often very harmful if you are not careful. As previously discussed, in modern manufacturing, there are often many Product CTQ specifications that are measured throughout the production process. Also, during final tests, many characteristics and CTQs are measured as they try to assess whether the product meets all functional and parametric specifications. Trying to implement a Maverick Program on many specifications within production or at final test was just impractical. Whenever I saw Maverick Programs implemented on specific or multiple CTQs either during production or at final test, it caused the very behaviors that rendered SPC programs ineffective - *competing priorities*. Indeed, the most frequent behavior was to ignore the SPC charts and only focus on whether a device or a lot was found to be a Maverick to determine if it could be shipped. Process Control and stability were no longer a priority. In addition, it led to countermeasures like adding specification limits to control charts. These behaviors triggered the *"domino effect"* rendering SPC programs ineffective. These competing priorities made it very difficult and costly to ship products.

13.2. Maverick Lot Program using Final Test Yield

JEDEC JESD50C allows multiple "Maverick" techniques. However, there was one Maverick technique that I used that was very effective; and, once customers understood it, they were pleased to accept it as a suitable "Ship-to-Control" method. It finds the right balance at detecting Maverick Lots while allowing effective SPC Programs to co-exist. It is based on techniques allowed by JEDEC JESD50C.

JESD50C allows the use of Final Test Yield (FT) monitoring and Part Average Testing (PAT) as key metrics, but they serve different purposes in outlier management. Final Test Yield targets Maverick Lots, while Part Average Testing targets Maverick Units. Table 34 provides a comparison of the 2 techniques.

Table 34 Final Test vs. Part Average Testing

Aspect	Final Test Yield	Part Average Testing (PAT)
Definition	• Measures the percentage of devices in a lot that pass functional and other datasheet specifications at final test.	• Compare each device's parametric values (e.g., leakage, timing) against the statistical average of the lot.
Application	• Applies only on Final Test Yield data.	• Applies to any CTQ specification at any time throughout production.
Focus	• Lot-level performance – is the overall yield consistent with historical norms?	• Device-level consistency – does each unit behave like the rest of the population?
Detection Method	• Yield shifts, abnormal pass/fail ratios, sudden drops in lot yield.	• Statistical outlier detection (e.g., >3σ from mean, abnormal distribution tails).
Risk Addressed	• Identifies "Maverick" lots that affects many devices.	• Identifies "Maverick" units (devices) that pass specs but are reliability risks due to abnormal behavior.
Example	• A lot normally yields 92%, but one batch drops to 77% – the suspect lot is identified as a "Maverick Lot."	• Most devices have leakage ~2-3 μA, but a few measure 9-10 μA – flagged as outliers even though they pass the ≤10 μA spec. • The offending unit(s) are identified as "Maverick Units."
Disposition	• Maverick lots may be quarantined, reworked (if allowed), retested, or scrapped.	• Outlier units (devices) are segregated, dispositioned, or eliminated before shipment.
Root Cause Investigation	• Maverick Lot root cause is typically easier to find that Maverick Unit root cause.	• Maverick unit root cause is typically harder to find than Maverick lot root cause. • As a result, investigations are often ignored, the units are just scrapped.
Effectiveness	• Frequent yield improvements occur because of effective investigations.	• Yields are rarely improved because of incomplete investigations.

Both these techniques work effectively at detecting outliers (Mavericks). But Part Average Testing (PAT) has a key disadvantage when production lots consist of many production units as modern products have a plethora of applicable product specifications while being manufactured! Attempting to implement these techniques on *all* key product specifications within a factory is impractical. It will either cause production delays (nothing will ship); or they will introduce other requirements that start the *"domino effect"* which as we now know will cause ineffective control charts. My suggestion is if Maverick outlier units that are fully functional and within specifications really have a higher reliability risk, try recalculating the specifications using the techniques described in Chapter 11 to re-adjust the acceptable distribution.

13.3. Final Test Yield Maverick Lot using I/MR Control Chart

Monitoring Final Test Yield of a production lot using an Individuals/Moving Range Control chart (i.e., a single production lot consists of multiple units) was my most effective Maverick Lot Program. Each lot typically undergoes a Final Test to determine how many units meet specified criteria, which includes both specification and functional testing. In the absence of a Maverick Lot Program, units that pass the Final Test are immediately approved for shipment. However, when a Maverick Lot Program is implemented, the organization monitors the Final Test Yield for each lot as an important performance metric. If the Final Test Yield is within control (IC), the passing production units with the lots are considered acceptable for shipment. However, if the Final Test Yield is considered Out of Control (OOC), the production lot is contained for disposition, and a root cause investigation is launched.

So how do you set up a Maverick Lot system using individuals and moving range (I/MR) control chart? Apply the control chart formulas described in Table 4 Key Formulas for Individuals, Moving Range Chart to the Final Test Yield measured for at least 30 production lots to establish the control limits for both the Individuals Chart and the Moving Range Chart. Since this control chart type is based on two consecutive individual lot Final Test Yield measurements, the moving range subgroup size is effectively 2, and the constants are always $E_2 = 2.66$, $d_2 = 1.128$, $D_3 = 0$ and $D_4 = 3.268$. Each individual lot Final Test Yield is compared against the control limits.

Figure 31 is an example of a Final Test Yield Maverick Lot control chart that I routinely used in the semiconductor industry. In this case, each production lot consists of 30 wafers, where each wafer may contain 500 computer chips - totaling 15,000 chips per lot. Chips that fail the Final Test are identified and excluded from packaging and shipping. This control chart shows an example where the Final Test yield is stable and in-control using the Individual/Moving Range control chart technique. So, all passing units within each lot were allowed to be shipped. No Maverick Lots are detected.

Figure 31 Example of Final Test Yield Maverick Lot Control Chart

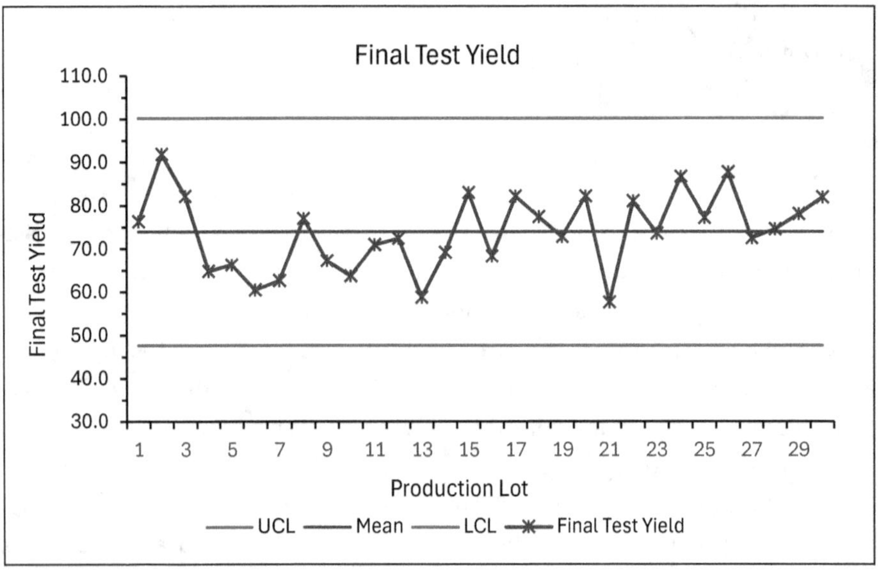

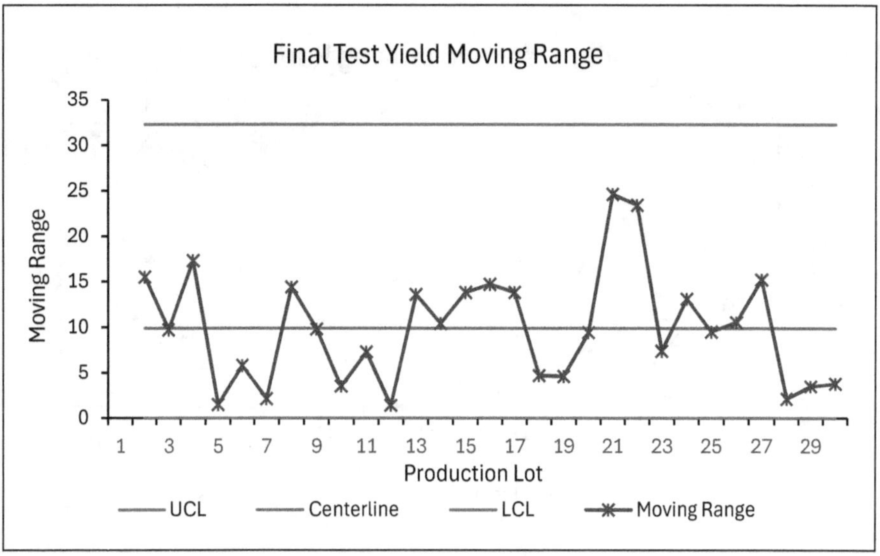

Figure 32 is example of a Final Test Yield Maverick Lot trend chart that exhibits 3 OOC events on the Individuals chart as low yielding lots.

Figure 32 Example of Final Test Yield Maverick Lot with OOC Events

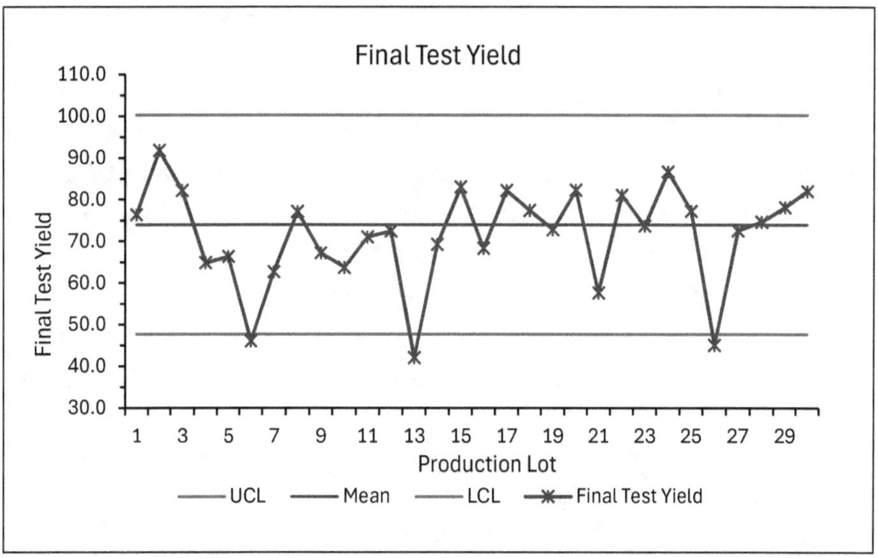

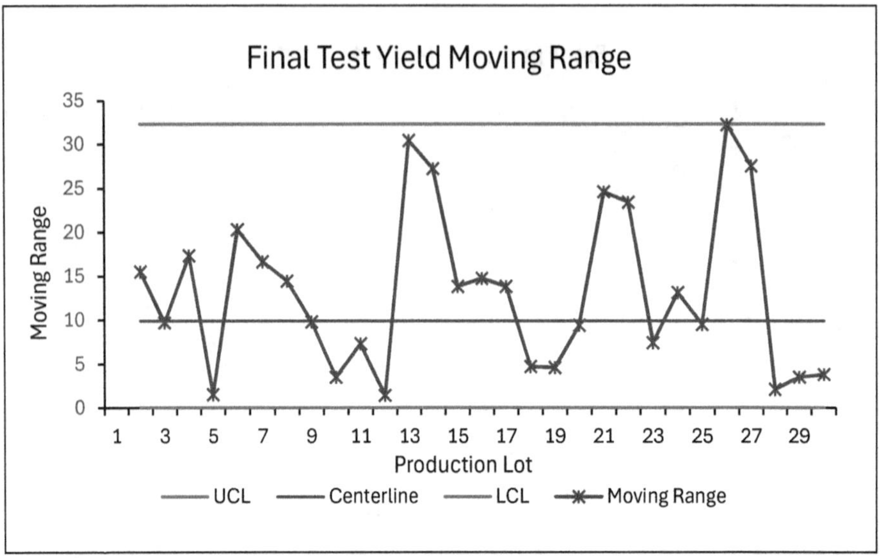

This Maverick Lot Program integrates Individuals/Moving Range (I/MR) control chart principles and is designed to detect and contain statistical outliers at the lot level. Unlike traditional individual specification-based screening, this approach leverages statistical process control (SPC) principles on the combination of final specification and functional tests to identify outlier production lots. Containing the production lot and launching investigations should reveal the assignable causes. Lots are flagged not only when individual Final Test Yield exceed control limits, but also when the moving range between consecutive measurements indicates instability.

SPECIAL NOTE: If average yield causes the individual chart's UCL to be exceeded, be sure to investigate why yield is unusually high. My experience revealed that successful investigations often uncovered testing errors or lead to lasting improvements in production yield.

Automation plays a central role in Maverick Lot Programs. When integrated with the manufacturing execution systems (MES) or quality management systems (QMS), the program can automatically screen lots prior to release, apply hold on production lots when they display an OOC event; and launch investigations. This ensures consistent enforcement of quality rules and rapid response to emerging issues.

13.4. Maverick Lot Protection

As suggested earlier in this chapter, there are other situations where I found this Maverick Program to be very effective: first, for organizations that were beginning to launch SPC programs; and, second for those organizations that failed SPC Program Audits as described in Chapter 12. This could be both within your organization and/or with your suppliers.

In these situations, I would quickly focus on implementing a Maverick Lot Program using Final Test Yield to ***protect*** the organization from releasing products that have a higher probability of warranty claims and/or reliability failures while additional efforts were used to upgrade their SPC programs using methods described in this book and address the findings from my audit. Simply put, it was much easier and quicker to implement a Maverick Lot Program on Final Test Yield than it was to deploy an effective SPC Program across a whole factory that failed one of my audits.

LET'S EXPLORE A REAL EXAMPLE

Upon joining one of the companies I worked for, I managed the Supplier Quality Engineering (SQE) Department responsible for managing our key suppliers. I quickly noticed two key offshore suppliers (call them Suppler A and Supplier B) in Asia had elevated levels of customer complaints regarding both product returns and warranty claims (most were reliability failures). I quicky organized my team to conduct supplier audits, including SPC audits, using the techniques described in this book. I conducted the SPC audits, as my techniques were new to my new company as well. As you might expect, they both failed my SPC audit. I also estimated that it would be more than a year to coach them on how to upgrade and improve their SPC program. The company didn't want to wait that long for improvements, so to protect my new company, I suggested the quick implementation of a Maverick Lot Program on Final Test Yield, which they accepted along with my continued follow-up to improve their SPC Programs. Let's look at what happened.

Fortunately, Supplier A was under new contract negotiations. I was able to get the Maverick Lot Program specified as a Lot Acceptance criteria. Whether the Final Test Yield exceeded the UCL or was below the LCL, the Lot would be quarantined, and an investigation would be launched. Further, our SQEs were part of the final Material Review Board (MRB) decision making team, so OOC lots could not be shipped without our permission. Unfortunately, Supplier B already had contracts in place that specified all lots with Final Test Yield >50% could automatically be shipped to us independent of whether it was an OOC event. Although they agreed to implement the Maverick Lot Program, they would not agree to using it as containment/shipping criteria. They wanted to use it internally only. However, both suppliers agreed that I could provide additional SPC Training to help them address my SPC Audit findings. So, what happened?

Supplier A initially had an average First Time Yield (FTY) of around ~74% when control charts were set up, and suitable upper and lower control limits (UCL/LCL) were established. Together, we examined out-of-control investigations and determined the disposition of lots. They approached these OOC events with strong commitment and thoroughness. Within one year, Supplier A increased their FTY to over 92%, successfully passing a second Statistical Process Control (SPC) audit by addressing all findings. As a result, customer complaints dropped significantly and warranty claims ceased. Both companies experienced considerable financial gains from these improvements.

For Supplier B, the initial average Final Test Yield (FTY) during control chart setup was about ~60%, with proper upper and lower control limits (UCL/LCL) established. I performed a second Statistical Process Control (SPC) audit six months later and noticed that while some issues had been resolved, others remained (especially competing priorities), resulting in another failed audit for Supplier B. Further, their FTY did not improve, and I noted that their documentation of investigations into OOC FTY incidents was inadequate. Although customer complaints and warranty claims decreased, this improvement was due to my company's efforts rather than Supplier B's. Because we were monitoring FTY, we introduced a Maverick Lot Program within our company only to determine lot disposition, since we could not carry out real-time investigations. As a result, many OOC low yielding production lots were not shipped. So even though our return costs were lower, overall, we had increased manufacturing costs because we still had to pay Supplier B for OOC lots that they were allowed to ship to us.

Well, a year later our contract for Supplier B was up for renewal. This time we had more leverage. We demanded this time the same Maverick Lot Program as was implemented with Supplier A. They continued to resist until we showed them the FTY Trend chart of their competitor, the financial benefits we both got; and indicated that without progress we would begin moving some of our products to Supplier A. Well, they finally agreed. One year later - Supplier B FTY achieved ~90%! They finally passed my SPC Audit and both companies benefited financially.

The results of these 2 suppliers had even wider consequences. In both cases, all companies benefited financially. Since I provided much of the SPC improvement training, using the principles described in this book - competing priorities disappeared at both suppliers. As you can imagine, many of my original SPC audits findings highlighted existing competing priorities. As they started addressing findings, eliminating existing competing priorities, and adopting other techniques described in this book, their SPC programs became more effective, and their processes became more stable. As a result, their process capabilities improved leading to more FTY improvements.

My company also saw these results, which helped me further improve our SPC Programs - a big win for all involved. Competing priorities began disappearing, our yields and process capabilities also improved. Further, my company began targeting the automotive market - which had "Ship-to-Control" requirements. The strategy of having a FTY Maverick Lot Program coordinated with (versus competing with) an SPC Program following the principles described in this book was widely accepted by all customers with SPC and "Ship-to-Control" requirements!

RECAP

A Maverick Lot Program can serve as a robust tool for fulfilling "Ship-to-Control" requirements, especially in sectors like automotive, aerospace, and semiconductors where the demand for zero defects is high. It also provides protection for companies while improvements to SPC programs, as outlined in this book, are underway and other priorities are being addressed.

Success depends largely on your implementation strategy. Although many methods for identifying Mavericks during factory production have been documented, they often end up causing more problems than they solve. Techniques focusing on individual units, such as Part Average Testing (PAT) based on certain parameters or specifications, frequently trigger *"domino effect"* practices that undermine the effectiveness of SPC Programs and increase operational costs.

From my experience, the most effective approach involved using a Final Test Yield Maverick Lot Program with an Individuals/Moving Range Control Chart. This method successfully met both internal and supplier requirements, as well as customer "Ship-To-Control" expectations. Additionally, I found it valuable for safeguarding our company during the initial rollout of SPC programs or when improving programs that had failed SPC audits, applying the techniques discussed in this book. It integrates statistical detection, segregation, and elimination of abnormal products across fabrication, assembly, test, packaging, and shipping. This ensures higher reliability, lower defect rates, and compliance with industry expectations.

14. EPILOGUE: THE JOURNEY TO EFFECTIVE SPC - A LASTING PROMISE

As we reach the end of this journey through the proven methods of effective Statistical Process Control, it is worth pausing to reflect on what has truly been accomplished – and what lies ahead for every reader, practitioner, and leader who seeks to transform their organization's approach to quality, reliability, and process improvement.

Throughout this book, we have explored not just the mechanics of SPC charts, but the deeper challenges and opportunities that arise when theory meets the realities of manufacturing and supply chain environments. We have seen how competing priorities – between product and process control – can undermine even the best intentions, and how common pitfalls in chart setup, sampling, and action planning can trigger the *"domino effect"* that leads to ineffective, compliance-driven programs.

But we have also learned that these challenges are not insurmountable. By embracing the principles of *effective* sample size, rigorous qualification, and the separation of OCAPs and OSAPs, organizations can move beyond "SPC for Show" to "SPC for Improvement" and realize the true promise of process stability, reduced variability, and continuous improvement i.e., the road to Operational Excellence! The case studies and real-world examples shared here are not just stories – they are evidence that change is possible, and that the benefits are tangible: lower scrap rates, higher yields, fewer customer complaints, and stronger market positions.

The journey does not end with the release of a qualified control chart or the passing of an audit. *Effective* SPC is a living discipline, one that demands ongoing vigilance, learning, and adaptation. As processes evolve, as new equipment is introduced, and as customer requirements shift, so too must our methods and mindsets. The tools and scorecards provided in these chapters are meant to empower you – not just to assess and improve your own SPC programs, but to drive a culture of accountability and excellence throughout your organization and supply base.

If there is one lasting message to carry forward, it is this: *Effective* SPC is not a destination, but a journey. It is a commitment to root cause analysis, to data-driven decision making, and to the relentless pursuit of process stability and capability. It is a promise – to yourself, your team, your customers, and your company – that quality is not just a requirement, but a competitive advantage.

May the methods, insights, and stories in this book serve as your guide, your reference, and your inspiration. Whether you are an engineer, auditor, manager, or executive, you have the power to transform your SPC program from a checkbox on a compliance list to a driver of real, lasting improvement. The journey is ongoing – and the rewards are well worth the effort.

Thank you for joining me on this path. Now, go forth and make your SPC truly effective!

15. APPENDICES

15.1. Table of Constants for X-Bar, R-Charts

Table 35 Constants for X-Bar, R Charts

Subgroup Size	A_2	d_2	D_3	D_4
2	1.880	1.128	-----	3.268
3	1.023	1.693	-----	2.574
4	0.729	2.059	-----	2.282
5	0.577	2.326	-----	2.114
6	0.483	2.534	-----	2.004
7	0.419	2.704	0.076	1.924
8	0.373	2.847	0.136	1.864
9	0.337	2.970	0.184	1.816
10	0.308	3.078	0.223	1.777
11	0.285	3.173	0.256	1.744
12	0.266	3.258	0.283	1.717
13	0.249	3.336	0.307	1.693
14	0.235	3.407	0.328	1.672
15	0.223	3.472	0.347	1.653
16	0.212	3.532	0.363	1.637
17	0.203	3.588	0.378	1.622
18	0.194	3.64	0.391	1.608
19	0.187	3.689	0.403	1.597
20	0.18	3.735	0.415	1.585
21	0.173	3.778	0.425	1.575
22	0.167	3.819	0.434	1.566
23	0.162	3.858	0.443	1.557
24	0.157	3.895	0.451	1.548

15.2. Table of Constants for X-Bar, s-Charts

Table 36 Constants for X-Bar, s Charts

Subgroup Size	A_3	B_3	B_4	C_4
2	2.659	-----	3.267	0.7979
3	1.954	-----	2.568	0.8862
4	1.628	-----	2.266	0.9213
5	1.427	-----	2.089	0.9400
6	1.287	0.030	1.970	0.9515
7	1.182	0.118	1.882	0.9594
8	1.099	0.185	1.815	0.9650
9	1.032	0.239	1.761	0.9693
10	0.975	0.284	1.716	0.9727
11	0.927	0.321	1.679	0.9754
12	0.886	0.354	1.646	0.9776
13	0.850	0.382	1.618	0.9794
14	0.817	0.406	1.594	0.9810
15	0.789	0.428	1.572	0.9823
16	0.763	0.448	1.552	0.9835
17	0.739	0.466	1.534	0.9845
18	0.718	0.482	1.518	0.9854
19	0.698	0.497	1.503	0.9862
20	0.680	0.510	1.490	0.9869
21	0.663	0.523	1.477	0.9876
22	0.647	0.534	1.466	0.9882
23	0.638	0.545	1.455	0.9887
24	0.619	0.555	1.445	0.9892
25	0.606	0.565	1.435	0.9896

15.3. Dataset with Samples that are Random and Independent

Process Run	Sample 1	Sample 2	Sample 3	Sample 4	Sample 5	Sample 6	Run Average
1	92.0	106.6	103.7	87.4	93.8	99.6	97.2
2	95.5	100.7	106.2	87.9	95.3	107.8	98.9
3	103.1	102.8	105.0	95.0	100.9	103.7	101.8
4	95.9	87.9	103.0	103.0	100.7	93.1	97.2
5	100.2	102.3	107.8	98.5	109.0	105.3	103.8
6	92.5	96.4	99.9	104.0	106.2	99.4	99.8
7	98.3	97.6	105.5	92.3	97.1	105.0	99.3
8	96.2	102.2	103.6	105.4	94.1	101.4	100.5
9	105.3	99.4	108.2	95.9	101.0	103.7	102.2
10	98.5	99.7	107.8	100.8	107.0	91.7	100.9
11	101.4	100.2	103.9	104.6	97.6	92.8	100.1
12	100.1	97.6	101.6	95.6	109.6	104.3	101.5
13	94.4	91.2	99.8	89.1	108.2	99.3	97.0
14	93.0	95.6	108.8	105.4	92.6	109.6	100.8
15	100.9	93.9	105.8	99.0	106.2	98.4	100.7
16	106.7	99.7	93.7	93.8	104.7	104.6	100.6
17	102.3	95.1	101.8	103.2	97.3	105.3	100.8
18	95.6	103.4	91.0	96.8	101.3	102.1	98.4
19	101.4	100.3	104.6	93.8	99.8	106.2	101.0
20	95.4	103.6	86.2	108.6	91.9	97.3	97.2
21	109.3	100.4	105.4	104.1	101.2	99.4	103.3
22	97.2	96.1	98.7	98.1	95.0	104.6	98.3
23	99.7	90.5	105.3	97.8	90.4	99.3	97.2
24	90.0	97.6	107.6	103.5	101.6	104.6	100.8
25	100.8	95.0	110.4	103.3	100.4	105.1	102.5
26	94.6	90.6	98.4	106.9	94.5	97.4	97.1
27	89.8	97.4	102.5	101.4	102.7	91.4	97.5
28	108.8	97.7	102.9	107.4	99.7	99.0	102.6
29	99.1	101.1	100.9	106.6	102.2	100.6	101.8
30	98.8	100.5	91.9	105.5	104.0	100.3	100.2

Average of Process Run Averages:	100.03
Average of All Individual Measurements:	100.03
Standard Deviation of Process Run Averages ($\sigma_{\bar{X}}$):	2.03
Standard Deviation of All Individual Measurements (σ_i):	4.99
Calculated (n_{eff}):	6.04
Actual Sample Size:	6

15.4. Dataset with Correlated Samples

Process Run	Sample 1	Sample 2	Sample 3	Sample 4	Sample 5	Sample 6	Run Average
1	95.5	113.8	99.6	109.1	94.5	104.9	102.9
2	103.7	91.2	99.7	88.3	100.3	94.9	96.4
3	94.6	116.0	103.5	101.7	93.6	103.3	102.1
4	107.7	99.8	104.9	106.3	100.0	94.5	102.2
5	103.7	88.2	98.5	90.6	103.7	102.2	97.8
6	101.7	102.9	112.3	116.9	104.7	101.8	106.7
7	107.1	100.1	94.1	100.8	101.9	97.3	100.2
8	100.7	115.6	104.3	101.8	96.7	102.9	103.7
9	109.3	97.4	113.8	92.5	105.3	98.3	102.8
10	109.3	93.2	96.4	92.6	104.8	91.8	98.0
11	95.8	88.3	83.8	96.6	96.3	81.7	90.4
12	99.5	90.9	96.5	80.5	101.9	83.3	92.1
13	102.9	100.7	103.7	108.1	96.4	103.4	102.5
14	98.2	112.6	99.3	115.1	104.4	110.3	106.7
15	103.3	96.7	105.0	99.9	98.3	88.9	98.7
16	92.0	80.0	96.9	95.8	103.7	87.9	92.7
17	99.4	95.7	96.8	78.9	95.5	91.1	92.9
18	89.1	107.2	97.3	102.4	106.2	102.8	100.8
19	100.8	81.8	102.5	87.5	107.7	79.1	93.2
20	95.7	102.0	93.6	99.6	100.2	109.7	100.1
21	100.1	85.8	98.0	86.2	92.8	79.9	90.4
22	97.5	93.2	99.0	87.1	98.1	91.3	94.4
23	92.1	101.4	93.5	88.8	94.8	95.4	94.3
24	96.8	99.5	102.1	108.9	107.4	105.7	103.4
25	103.4	100.5	101.8	104.4	96.9	108.2	102.5
26	102.9	90.0	111.7	91.0	101.7	94.4	98.6
27	96.8	83.5	97.5	88.5	92.8	84.6	90.6
28	91.7	91.9	90.7	89.3	100.9	86.1	91.8
29	109.0	102.1	103.6	111.4	97.0	103.4	104.4
30	93.9	96.3	100.8	104.5	99.2	101.3	99.3

Average of Process Run Averages:	98.43
Average of All Individual Measurements:	98.43
Standard Deviation of Process Run Averages ($\sigma_{\bar{X}}$):	5.05
Standard Deviation of All Individual Measurements (σ_i):	7.5
Calculated (n_{eff}):	2.21
Actual Sample Size:	6

15.5. Individual SPC Chart Score Card

Auditor:		**Date:**	
CTQ Name:		**Product CTQ:**	☐
Chart Owner:		**Process CTQ:**	☐
Process Engineer:		**Starting Score**	**100**

	Criteria	**Possible Adjustment**	**Actual Adjustment**	**Score**
1	Control Limits Appear Unreasonable ☐ Appear too tight ☐ Appear too wide ☐ Other_______________	-10		
2	Specification Limits Plotted on Control Chart	-10		
3	Specification Limits are defined for a Process CTQ	-10		
4	Control Chart plots Product Lots versus Process Runs (exception, no deduction if 1 production lot = 1 process run)	-10		
5	OCAPs are not clearly documented	-10		
6	OCAPs specify investigation of the OOC Root Cause	+20		
7	OCAPs specify consistent adjustment	+10		
8	OCAPs allow inconsistent adjustments	-10		
9	OCAP calls for adjustment when OOS sample detected	-10		
10	OSAPs & OCAPs clearly different	+25		
11	OSAPs for Product CTQs are not clearly documented	-10		
12	OSAP calls for adjustment when OOS sample detected	-10		
13	OSAPs calls for containment, product disposition or rework the OOS material, no process adjustments are allowed	+20		
14	Able to Verify OCAP execution for all OOC events	+25		
15	Cannot verify OCAP execution for all OOC events	-10		
			Control Chart Final Score:	

Key Definitions	
OCAP = Out of Control Action Plan OSAP = Out of Specification Action Plan CTQ = Critical To Quality	OOC = Out Of Control OOS = Out Of Specification

15.6. SPC Program Scorecard

Control Chart	CTQ Name	Control Chart Owner	Department Owner	Score
1				
2				
3				
4				
5				
6				
7				
8				
9				
10				
11				
12				
13				
14				
15				
16				
17				
18				
19				
20				
			Average Control Chart Score:	

INDEX

E

F

G

I

J

L

M

N

O

P

Q

R

S

T

U

V

W

X

ABOUT THE AUTHOR

Steven F. Bergeron brings over 45 years of experience from progressive organizations across a variety of industries. After more than three decades as a Quality & Reliability Assurance Executive, he established QLRFocus, LLC (Quality Lean and Reliability Focus) to support startups, small enterprises, and global corporations in advancing operational excellence through specialized Quality, Lean, and Reliability strategies. His comprehensive services span multiple sectors, emphasizing effective standards compliance, process improvement, quality management systems, customer complaint management, supplier management, and data-driven reliability engineering. Mr. Bergeron's expertise includes international QMS standards (ISO, IATF, AS), Six Sigma, SPC, Lean transformations, and advanced product qualification (APQP) methodologies.

His Quality discipline proficiencies include:

- Implementing *effective* Statistical Process Control (SPC) methods using methods described in his book. His methods include best practices for addressing "Ship-to-Control" and Maverick programs.
- Advising on the establishment of Quality Management Systems for startups and small businesses.
- Consulting on the transformation of legacy or underperforming quality systems into high-performing, customer focused systems that deliver better business performance and higher customer satisfaction.
- Providing training, gap assessments, auditing, and program management to achieve compliance with ISO 9001, ISO/TS16949, AS 9100, ISO 13485, and other key Quality Standards.
- Providing APQP training and implementation, featuring workshops on PPAP, DFMEA, PFMEA, MSA, and SPC.
- Improving complaint management systems to boost customer satisfaction and product quality.
- Managing comprehensive Supplier Management solutions such as assessments, audits, qualifying suppliers, and performance tracking with Supplier Report Cards, SCARs, and other various Key Performance Indicators (KPIs).

- Coaching on the implementation of Lean Principles within Quality Management Systems and broader business processes.
- Delivering Reliability Engineering Services such as Weibull analysis, accelerated testing, degradation analysis, and MTBF/MTTF/FIT rate calculations.
- Extensive experience in experimental design and analysis (DOE).

Driven by scientific curiosity and a passion for solving complex problems, Mr. Bergeron's approach stems from his educational background. He holds a Bachelor of Science in Chemistry from Georgetown University and a Master's in Chemistry from University of California, Santa Barbara (UCSB). His learning and experiences at these universities sharpened his analytical skills and innovative thinking. He later strengthened his leadership abilities at the AEA Stanford Executive Institute.

During his career, the American Society for Quality (ASQ) certified Mr. Bergeron as both a Quality Engineer and later as a Quality Manager. OMNEX recognized his proficiency in key automotive disciplines, including APQP, PPAP, DMEA, DVP&R, and VDA6.3 process auditing. Reliasoft certified his expertise in Accelerated Life Testing Analysis and Reliability Engineering & Life Data Analysis. Motorola University awarded him a Six Sigma Black Belt, while Pall Corporation qualified him as a Lead Facilitator in Lean Manufacturing. Additionally, he held auditor certifications for leading QMS standards such as ISO 9001, IATF 16949, and AS 9100 from multiple organizations.

Mr. Bergeron began his career at IBM Corporation and Motorola, Inc., pioneering advanced experimental design and statistical process control as both a Process Engineer and Process Engineering Manager. His ability to translate technical knowledge into practical, results-oriented strategies propelled him into senior leadership roles. He has served as a Quality Executive at firms including Digital Equipment Corporation, Unitrode Corporation, Semtech Corporation, Conexant Systems, Inc., STEC, Inc., Pall Corporation, and Cree, Inc. As Chief Quality Officer, his responsibilities spanned Quality Assurance, Quality Control, Customer Complaint Management, Reliability Assurance, Supplier Quality Management, Failure Analysis, and Chemical Compliance Management departments.

Mr. Bergeron's career has been defined by a commitment to collaboration, ongoing learning, and evidence-based decision-making.

Renowned for his direct approach to database development, workflow design, and mentorship during organizational transitions, he is valued as accessible, insightful, and dedicated to supporting others' success. He excels in orchestrating transformational initiatives often spanning competing departmental priorities, adeptly managing change, motivating cross-functional teams, and transforming challenges into opportunities. Whether conducting global quality audits, coaching teams, or developing training programs, he demonstrated empathy, adaptability, and an unwavering pursuit of operational excellence.

He was recognized by "Who's Who in Quality" and was an editorial advisor for selected articles in the "The Quality Observer." He was twice appointed by the U.S. Department of Commerce to the National Quality Award Board of Examiners as a Malcolm Baldrige Examiner.

Mr. Bergeron guided startups, small businesses, and multinational enterprises on standards compliance, process optimization, supplier management, and reliability engineering. He remains committed to advancing quality systems and fostering cultures of operational excellence.

Mr. Bergeron is widely regarded by his colleagues as an expert in the assessment, design, and application of Statistical Process Control (SPC) techniques. He has successfully led many SPC initiatives in diverse organizations and industries, which motivated him to author this book.

www.ingramcontent.com/pod-product-compliance
Lightning Source LLC
LaVergne TN
LVHW022217200926
848481LV00006B/2001